Dropshipping

Ultimate Guide to Private Label,
Retail Arbitrage and finding
Profitable Products

*(Make Money Online selling Physical Products
Without Having an Inventory)*

Chris Perkins

Published By **Regina Loviusher**

Chris Perkins

Dropshipping: Ultimate Guide to Private Label, Retail Arbitrage and finding Profitable Products (Make Money Online selling Physical Products Without Having an Inventory)

ISBN   978-1-77485-704-5

Legal & Disclaimer

The information contained in this ebook is not designed to replace or take the place of any form of medicine or professional medical advice. The information in this ebook has been provided for educational & entertainment purposes only.

The information contained in this book has been compiled from sources deemed reliable, and it is accurate to the best of the Author's knowledge; however, the Author cannot guarantee its accuracy and validity and cannot be held liable for any errors or omissions. Changes are periodically made to this book. You must consult your doctor or get professional medical advice before using any of the

TABLE OF CONTENTS

Chapter 1: Types Of Dropshipping

There are various types of dropshipping. The type of dropshipping you need to sign up to will be contingent on your needs. The two main types of dropshipping are blind and straight dropshipping. If you don't care about revealing the name of your supplier to your customers and customers, then straight dropshipping could be the best option for you, but should you need to mark your products under your own name, then the best choice would be blind dropshipping.

Straight dropshipping

This is the kind of dropshipping that you can choose if you don't want to conceal your supplier name from your customers. This kind of dropshipping provides the suppliers the opportunity to market their own brand rather than the merchant. If you are looking to establish a business that is serious then this isn't the right type of dropshipping option for you, but If all you're interested in is handing over customer service or warranty service, as well as other unorthodox conditions to the supplier, you could consider using this dropshipping method. When you put this kind of dropshipping in place the vendor might also prefer to use his own label and may also refuse to brand the product with the name of the seller.

Dropshipping without a label

In dropshipping that is blind the seller acts as a substitute for the seller. The product is shipped under the name of the vendor and conceals his identity from customers. The name of the merchant is mentioned on the shipping address and is able to claim complete liability of the item. This is the kind of dropshipping suggested for those looking to build lasting businesses. It stops the seller in using the dropshipping method to promote his company and permits the seller to market his own business. It is the ideal method of dropshipping to get your customers to believe that you are the owner of the product and is typically the most utilized method of dropshipping. As you'd imagine, the retailer takes all responsibility for any problem that arises in relation to products, warranties, and customer service. The merchant may claim that the vendor will be responsible for the problems caused by the delivery delay, however, the customer holds the supplier responsible. It is clear that the customers know little about suppliers, and, in the real sense, he's not within the chain of production. This is a complex and demanding model for business that will bring more challenges to handle but it could make a significant difference in the long run for the company.

Benefits of dropshipping

The rapid growth of dropshipping clearly puts the way to its many benefits. The benefits of dropshipping can be summarized in a few sentences.

Start-up capital is not required. One of the main reasons dropshipping is now a popular business is the fact that sellers do not require be able to raise any money to venture into dropshipping. Traditional retailing involves investing perhaps thousands of dollars the beginning, while dropshipping lets you start your business without investing any capital. It is possible due to the fact that you buy products from sellers using the same amount of money that is that consumers pay. The merchant would have completed the purchases prior to making the purchase, which allows dropshipping appear as an ideal business model that is win-win and profitable for new businesses.

Expand your business easily It is easy to expand. You don't have to be tied to just one supplier when it comes to dropshipping. You can purchase your goods from as many different suppliers as you can. This allows you to broaden your offerings and niche, and also to provide more superior products and services for

4

your clients. Since the dropshippers manage the shipping process, the dropshipper can have all the time to concentrate on the most important areas of business as well as marketing. This allows the dropshipper to get the time needed to grow the company and add other products. To expand the business, the dropshipper typically requires a functional and robust website which can bring in a significant number of clients. Naturally, the company becomes more complicated as it expands, but because the dropshipper does not provide to the shipping process or the delivery of products, he is able to concentrate on the essential aspects of the business, no matter how complicated.

It is easy to get started The barrier to entry into the dropshipping industry is extremely minimal because it is easy to get started. Because dropshippers do not take care of the shipping and stocking costs, warehouse and inventory and also manages the inventory level and inventory, it's relatively simple and less complicated than a brick and mortar type of business. When you add it all up, the seller of a dropshipping enterprise has an overall cost that is relatively lower as compared to the counterparts who operate who retail bricks and mortar. It is possible to operate the entire dropshipping enterprise at home, and spend only a small amount on the equipment needed to stay online. Naturally, the costs and requirements for dropshipping will increase when the business expands and expands, but the costs remain low in comparison the bricks and mortar stores.

The benefits of dropshipping

The advantages of dropshipping can make it an appealing option, but it is important to evaluate the pros and disadvantages prior to attempting the business. The disadvantages and drawbacks of dropshipping are discussed.

The logistics are complex as company grows - Dropshipping logistics may become extremely complicated as the business grows. The majority of dropshipping merchants have multiple suppliers and their suppliers rely on several warehouses for the products. This makes it extremely difficult to manage inventory and could affect customer service. Inability to efficiently and effectively manage the logistics may lead to inefficiency of the tracking system for products delay in shipping and other issues that eventually result in poor customer experience. Although there are solutions to improve inventory and logistics however, certain systems might not deliver the level of efficiency needed for managing logistics.

A low barrier to entry It may appear to be an advantage, but in actuality, it's a disadvantage. Since it's easy to establish dropshipping businesses and does not require capital, almost anyone can establish his own company. This can mean increased competition, lower margins, and other disadvantages. The low margins can alarming. A business owner may choose to reduce the cost of his products to a minimum to get customers , but in the long run the customer service could suffer terribly. Since the providers are open for business, they may offer drop-shipping to any other company and you. This means that the competition is tight and stiff.

Insufficient uniqueness - Everyone is looking to dropship. This is why uniqueness is a major issue for dropshippers. A majority of the products of dropshippers are offered by other sellers and buyers are able to find another retailer to buy from. Dropshippers, however, go to the greatest lengths to search for distinctive products. However, since they are a very competitive market, their products will quickly lose their distinctiveness as others dropshippers will begin selling their products.

Common dropshipping mistakes

The new dropshipping merchants entering the market can fall victim to common mistakes in dropshipping which make it hard for them to expand their business and increase its size. To be successful as a dropshipper should learn from successful businesses and those who have tried business model and have succeeded. There are a few mistakes that you must avoid as merchant.

Backorder - A backorder is when a retailer places an order for a product that's at present not in the inventory of the supplier. Backorders in dropshipping can mar your business image. As a dropshipper pay prior to purchasing from the vendor. Backorders mean that the customer has completed their purchase, but you're unable to provide the item because it's not in stock at your supplier. To address the problem of backorder, it's best dropshipping businesses to make use of at least as many vendors as they can. This will allow customers to find a new supplier if one is not in stock.

Reactivity and poor customer service poor customer service can create a an extremely negative impression of your business to customers. To be successful in the highly droppingshipping industry it is essential to take every step to convince your customers that you're the top. This means going the extra the extra mile in terms of customer service. Value delivery must encompass all aspects of your company such as the product, customer service shipping, and so on. The ability to respond is a crucial component of client service. It is crucial to address customer's needs and concerns in the quickest time possible and give them accurate information regarding your services and products.

Insufficient or inadequate branding Branding is an essential aspect of dropshipping. If you're looking to get high returns from customers You must focus the importance of branding. Of course, you are able to make the item more attractive by putting your logo on the item however, you must ensure that you leave a lasting impression on your customers throughout the delivery process.

Customers should also be provided with your policy and contact details to strengthen the trust of your customers.

Insufficient system for cancelling or return items - Although it's great to have the provision that customers can purchase goods, you must also have a plan for cancellation and return of purchases. It is crucial to be prepared for the worst-case scenario. Your system must allow customers to cancel their orders quickly and easily so that customers can cancel orders, and receive their money back. This will help build customer confidence in your service and will help build your business.

Delivery delays - Delays in order processing are a common problem for dropshipping company. Customers are less loyal and confidence in your business when their order is unnecessarily delayed. In reality, dropshippers are in no control over the order procedure, but they can take steps to stop delays from occurring at all. One of the most important actions you can take to stop delays on orders is to make sure that your suppliers offer the most current inventory of their products every day. This can help let customers know what products are that are in stock, thereby avoiding having to order products that aren't available.

Additionally giving your customers tracker information for their purchases can aid in saving the day. They'll be able to rest in assurance knowing the condition of their goods at any time. It is possible to solve the problem of delayed orders by ensuring that you offer your customers high-quality customer service. Keep them informed of the status of their order at any moment and also help them keep an eye on the status of their orders.

Chapter 2: Setting Up Customer Service

It is crucial to provide excellent customer care cannot be overemphasized. This is the reason keep customers returning to you time and over. The majority of drop shippers believe outsourcing this work to their supplier. If they locate a good supplier, they don't need to worry about dealing with problems with customers.

As I stated earlier that even the top Tier 1 suppliers are likely to fail at times. It is essential to establish a procedure that reduces complaints from customers and handle any issues that might occur.

The importance of customer service

If you're not sure that customer service is crucial to your business Here are some figures to think about. Research has shown five-six percent of customers are unlikely to purchase any products from a company again when their customer service isn't good enough (Dropshipping in 2019,). 20 percent of them will leave negative reviews online and 25% of them will advise their friends to not use this company in the future. Furthermore 14% of them are required to share their experiences via social media. Personally, I think that number is somewhat low to me, considering the instant ability of people to write reviews.

Companies in the US suffer a loss of $83 billion every year due to bad customer service (Dropshipping 2019, 2019). What's more is that a negative customer experience is more likely get shared rather than a positive one. Therefore, the lesson to you to take away from this is that you need to build a solid customer service skill. Otherwise, you're at risk of making losses.

In spite of all this we must face the facts. It is likely that you want to set the dropshipping business in order to enhance the degree of freedom you enjoy in your daily routine. It's not possible to deal constantly with clients regardless of how important they may be. It is in your best interest to be able be prepared for frequent queries you'll be asked and to design your website in as to minimize the number of questions from customers.

Some are:

Can I receive a refund or a refund?

Where's my belongings?

That's all there is to it! Simple, right? Let's take a look at how to answer these issues in a way that is appropriate.

Refunds and Returns

If a buyer is disappointed with a item, they are likely to demand a return or exchange. They'll request an immediate refund. This is when you will have a decision to make. Many store owners discover that it's more profitable to deliver another item to the customer at no cost, instead of asking customers to return the item and then replace it.

Naturally, you'll want to avoid refunds at every opportunity. Therefore, send them an email and offer apologies for the difficulties you're having with them in the most polite manner. Don't reply with a template, instead, write in the manner that the person you are writing to. Offer to repair their defective item for free or provide them with credit at the store either as an upgrade, coupon.

It is likely that a client who is important to you for the rest of their entire life will usually agree to this. It's a pity for some customers to insist for a full return at all costs. In these cases, you can give them their money back and kiss them goodbye. The reason you wouldn't wish to make a refund to a customer right away, however, is because of the life-time worth (LTV) that you receive from a client. Simplyput, you want to maximize the revenue you get from your customers. This will reduce your cost of acquisition and advertising over the long term. The ability to build loyalty is essential to getting this.

Is There My Stuff?

This is fairly common when your shipping is internationally. Even if you post the shipping times in large bold letters on your site's homepage, so that it isn't ignored, customers will be able to message your about this. It's inevitable, honestly speaking.

Please respond in a professional manner by including their tracking code along with the URL where they can monitor their package. Include instructions in the email regarding how they can utilize that link and include the most current information which the link will display.

For instance, you could put something in writing, like "Your package arrived at XYZ and will arrive at your door at ABC." In most cases than not, you won't hear from them once more.

Customer Service Channels

Channels are simply a method of naming different ways through which your customers are able to contact you. The most commonly used method is via email, but you can always choose other methods. There are many options to choose from.

some of themit's dependent on you. In the meantime, let's examine a few channels more in depth.

Email

It is the most widely utilized channel. Its advantage of this channel is you do not need to respond instantly, which means you can plan your responses accordingly. The drawback is that the client will be more likely to become anxious if there's no response within the specified time frame.

There should be an Contact Us page at a minimum on your site. It is a good idea to put an FAQ page in a prominent position above the form. Do not ask for more details than you're required to. Some websites require the person's initial and last names as well as email address, telephone number or tracking ID, a fully written problem or other information. If you don't need a particular specific piece of information you don't need, don't ask for it. People are concerned to their personal information, and you need to behave accordingly.

Don't make the mistake of putting the contact form in a shady place within your FAQs page, as some large companies (e.g., Paypal) do. Make it more prominent. Remember, the more genuine you look more likely you will be to keep the client. Don't appear as a shrewd machine.

Live Chat

Live chat is a fantastic alternative that is growing in popularity each day. Chatbots let you respond to most queries from customers like tracking shipments. It is possible to customize it to search for the order details of your customer and provide them with an tracking ID and then update them immediately whenever the customer is online.

Custom queries can be revisited at a later time, and you can respond to them via email. Chatbots ease the burden on your shoulders and are becoming an acceptable method of interaction with customers.

Make sure you can identify the bot as a human and do not try to disguise a bot as a person. This could lead to backlash.

The only issue in chatbots are that a few people prefer to post crude messages or questions to harass the bot. In every good thing, there is a bad, and you'll have to face it. Intercom is an excellent service provider for chatbots. It is employed in dropshipping circles extensively.

It is possible to increase the brand's reach through your bot by giving it a meaningful name. For instance, TigerAir, a budget airline based in Australia is using Toby the Tiger on FB Messenger ("Toby the Chatbot" 2019.). You can send a message to Toby to receive information about your flight as well as reminders, restrictions, and other information. This will eliminate the need for humans to assist with trivial questions.

Social Media

I would not recommend this method unless you're a person who can commit entirely to this. Social media requires immediate response particularly on platforms such as Twitter as well as Facebook. It is a plus that people tend to stay your clients and also spread the word about you when you have a solid online presence.

Telephone Access

This is a traditional methodthat works, however it is effective. People like talking to someone in person, however the downside is that you must establish the infrastructure to handle this. This all costs money. If you're selling a costly product that has high margins, you must employ this method.

Reducing Demands

It's much better to prevent something than solving it. This is my grandmother's words to me. In the case of the customer experience, this phrase is very sensible. It is essential to organize your website in as to minimize the number of questions from customers. The simpler it is to for customers to find the answers on your site the more pleased the customers are. Here are three strategies to ensure you don't have to deal with customer complaints.

Frequently asked questions

You can make your own page or utilize an app or a plugin dependent on the platform you're using (more on this in the following section). No matter which option you pick the most important question you'll have now is "Which questions should be included to an FAQ?"

Here's how you can think about the issue. Sort your FAQs into the following categories:

The most frequently asked questions are recommended.

Orders in place

Placing orders

Delivery and shipping

Payments

Refunds and returns

After we have our primary categories in order, you can look into them in more depth. There are several methods to do this. The most effective method is to think of all the questions your customer is likely to confront. The simpler method is to go to another popular dropshipping website and repeat their concerns.

Always put your FAQ page on a page that is easy to locate. It should be as simple to use as you can. In spite of what you may see when you look at social platforms, we are pretty intelligent. If you are receiving inquiries that are answered in your FAQ page over and over, the problem lies with the placement and design of the page.

Tracking Page

It's a game changer in terms of decreasing customer inquiries. If your customers are able to input their tracking number into a form on your site and if you provide them with instant updates, they're less likely to contact you with a query. In the case of Shopify it is possible to use several excellent apps, including Trackr, Aftership, and Tracktor.

These applications allow you to notify your customers whenever they receive an update on their purchase.

Uncover the mysteries of shipping

This is the biggest problem for any dropshipping business. It is inevitable that shipping times will be delayed. In the time of Amazon Prime's next day shipping, customers aren't willing to wait for their items for more than a month. This is the reason we went through the filters before deciding on the right niche and product.

This is the reason I don't suggest purchasing goods from Aliexpress or any other site that is near China. In fact, it is better to locate your items locally in the country where you're selling. If you're selling your products in China or elsewhere, feel free to utilize Aliexpress. If not, look for suppliers who are located near to your customers. A great resource is Spocket. It can help you find suppliers within both the US or EU.

You'll still have to remove suppliers selectively , based on the criteria I've mentioned earlier however it will make life considerably simpler. Of obviously, if you follow the procedure I explained previously, you will not encounter any issues in identifying local suppliers. It's helpful to keep Spocket as an emergency backup.

Your shipping policy should be explicit and inform your customers by email of the length of time that their orders are expected to be processed. If you must ship your orders via Aliexpress and you are determined to making it happen, go with the option of e-packet. It is more expensive but is well worthwhile in terms of the shorter time to deliver.

Overall, you should ensure that your shipping policy is clear to avoid any problems. Great FAQs as well as tracking websites make this easy for customers grasp and ask questions.

Chapter 3: Drawbacks Of Dropshipping

Dropshipping is a business that offers a variety of opportunities for you to enjoy. It's a fantastic way to earn extra revenue in addition to your normal occupation. There is a broad variety of items to offer, and you can decide how much or much you want to offer and you're free to pick a schedule that will meet your busy schedule. However, there are some negatives to consider are faced when you enter the dropshipping market, and that is the reason only a handful of customers have made it through this type of business. Some of the major drawbacks that dropshipping can present are:

Sudden shortage in stock

As dropshippers, it's your responsibility to be up to date with the quantity of stock that is available at your source. Keep this information up-to-date within your store to let customers are aware of the moment the item is not in inventory or is not available. Sometimes, this is simple to achieve. But around the holidays or with a trending thing, it's difficult to stay on top of the latest trends.

If there's an unexpected shortage in the stocks of an item this could cause a problem for you. Customers might be disappointed by the inability to obtain that product immediately. If they've previously placed an order on the product, only to find that the item was not in inventory, this could create a problem for both you as well as the customer.

The best approach to address this is to find multiple sources for the same topic or at minimum, two or three suppliers who have similar issues. If your primary supplier is having a shortage of a certain area, and you've got potential customers who are interested, you have alternatives. If the products are identical, you can change suppliers and then return the item.

If you notice something different, you can reach out to the client and provide the alternative. You could also offer some extra incentives.

The customer service aspect is everything to you.

If a customer is unhappy about something, they'll not contact or write to the company. You are the head of the line for the business. As far as they are aware, you hold total control over your product. If something goes wrong, you have to handle the customer service issues yourself.

It can be tiring and sometimes difficult. It is your responsibility to answer any queries that the customer may have. It is your responsibility to respond to emails whenever you get comments, questions or complaints from a client. If there is to be exchanges or returns then you're the person who has to manage the entire process. As a human being who is busy, this could seem overwhelming and could increase your workload in certain situations.

You have less control of your personal business

Dropshipping is an excellent business idea to start. You can earn profits from the products other companies produce and you don't need to have stock or produce the product on your own. The downside is that you'll have only a little control over your business. The providers you select will have the greatest control over this kind of business. If you select the wrong supplier this could lead to the demise of your business.

As a dropshipper you offer items on the internet for different companies. You can register them for more than the company that they are registered with and then collect the revenue. Your customer will purchase from you, and the customer will pay you and place an order with the supplier. The supplier is in charge.

If you've chosen the best company to deal with, the entire procedure should be simple to handle. Once you have placed your order they'll make the item then deliver it to you to your customer, and they is sure to be satisfied. However, you have no say in this. The vendor could be able to deliver the item to the wrong address and orders may be misplaced. When it happens, you need to take care of the fall even though you have no control over it.

Potential problems with quality control

Because you aren't the creator of the product, and you do not use it, there might be issues with the quality that the item is manufactured. The company will typically attempt to be as efficient as they can, since that is how they earn money. If there are problems with quality control then you will be the one most affected. Customers will leave negative reviews, and there's not anything you can do since you don't make the product. There are some ways to ensure that you are providing top-quality products for your customers. When you're searching for a supplier look into the company. See the opinions of other drop-shippers who have to say about the services. Check to see whether there are any major problems with the business that you should be concerned about. If there are lots review negatives or issues,

Before you decide whether to offer an item or not you should consider purchasing it for yourself. In this way, you will gain an understanding of how the customer will feel when they place an order via yourself. You can find out what shipping process is, and then check with customer support and find out the process with the information you have available. Check this out every whenever you decide to sign up with a different provider for your company.

It is difficult to find products that can generate enough revenue

Drop shippers often have trouble having to find a product from which they are able to earn a good profit margin. There are a lot of businesses and suppliers who utilize this business model, however it is essential that you pick a product which can be sold for a profit. If you examine the cost of the vendor and the item is listed at $ 10 however, everyone else charges the price of 10.50 This is not the best product to market because you'll have to generate enough revenue to pay some thing worth your time.

Many businesses are similar which is why dropshipping has a negative reputation. It is crucial to be patient and not be rushed into the product you wish to use. The greater your margins, the more likely it will be for you to earn a revenue, and also the more worth your time. Don't spend your time earning only $0.50 0.50 for every product you sell. When you are working, socialize and pay fee for listing sites you'll lose cash. Choose products that perform as well as they can. If, for instance, you come across an item that costs the seller $50 however, other sellers are selling it online for $200 This is an item worth considering.

Provider errors

Sometimes, the vendor could commit an error on one of their orders. They could misinterpret an address with a different one and ship the item to the wrong location. They might send the incorrect item to one of their clients. They could also make a error that causes a problem for the customer.

If you're drop shipper, it can be difficult to convince the service provider to resolve this issue. A few of the top ones can help but the client could continue to be angry when something goes wrong in one of the orders. If this happens often it is possible to have several mistakes and poor reviews, and eventually, no buyer will be tempted to purchase from you again in the future.

It is better to choose a supplier that you can be confident in. One that is known for their accuracy in order and offering excellent customer service. Be aware that you represent the company. If you are selling the product, the client is likely to blame the seller for everything that go wrong. Even if you simply offer the product and then make the purchase then the customer will think that you're the one to blame for everything, and they will be able to be relieved of the anger and anger. Selecting a reliable service to care for your customers will be a major factor in your company's success.

There are many advantages for droppingshipping as a new venture however, there are some things to think about prior to starting. There are some responsibilities associated when you run a dropshipping business. While the potential for earning could be substantial but you'll need to put in hours and efforts to locate the perfect supplier. Someone that is able to provide excellent customer service. makes high-quality products, and gets orders from the most reliable customers. If you are able to do this then you will be able to avoid some of the disadvantages associated with drop shipping.

Chapter 4: Selecting An Area Of Interest

Okay, now is the time to learn how to start your own dropshipping company. It's not difficult at all due to the numerous online resources you can access.

To aid you in your journey, here's a list of the most essential things that you'll require for starting your dropshipping company:

A website shop to showcase your products you plan to sell

A reliable and reliable software that lets you add products from your suppliers to your website and track your purchases

A supplier or several suppliers that will be the source of your products you can sell

Software to assist you in increasing your online shop's online visibility

Picking the right niche

When you register to a website shop and begin searching for suppliers, it is essential to first select a niche set of items to offer. The subject matter you choose should be something you're passionate about, and have a good understanding of. It is important to tailor on your product lines to the area you are interested in, or your store will appear unorganized which makes it difficult to promote.

There are many reasons why you need a well-defined niche in Dropshipping

It lets you target your market - If you've decided to go with an area of interest it will be easier to figure out which areas to advertise your site by choosing a targeted market.

It lets you optimize your website to suit your needs. When you've identified an area of expertise it is much easier to optimize your online store for SEO. Without a niche and using only general phrases getting your website to the top of result pages is nearly impossible.

It lets you develop new ideas about it. If you're in a niche market, you will be able to quickly come up with new concepts that meet the requirements of your intended customers. Because your market is narrower when you have a niche and a niche market, finding solutions is simpler.

This will reduce the amount of competitors - If you are in a niche that is highly specific, area, the likelihood of being in a position to have a significant amount of competition are very low. A broad market is similar to an ocean of bloody hungry sharks. If the opportunity to sell arises and thousands of sellers are available, they will rush to the opportunity, which makes it difficult to stand out to get the sale. The existence of a niche reduces the number of vendors who you must fight against as well as those customers who reside within your area are typically extremely interested in the subject which makes them more likely to purchase.

Tips for Picking a Good Market

It is essential to be passionate about your field, as it takes many hours of work in order to turn a dropshipping enterprise profitable. If you're not interested in the products you sell, then you may easily be frustrated and give up. There are a few things you could consider when choosing an area of specialization:

Find products that have large profit margins. When it comes to dropshipping it is important to note that the effort to sell less expensive product as well as higher priced ones is basically identical. In both cases, you are putting all of the time to market and find potential customers, therefore it is best to offer higher priced products to have the chance to earn some money from every sale.

Choose suppliers with lower shipping charges - despite the fact that it will be the seller that will handle the shipment of the product however the cost will be passed on to the purchaser. That means, even though a product is priced at a an affordable price compared to your rivals, if shipping costs are higher than the price this will cause a loss to customers. Look for products that are reasonably priced to ship, as this gives you the opportunity to provide no-cost shipping options to clients. If the profit margin on the product is high enough, you can take on the shipping cost as a cost of business, and make some profit from the sale of the item.

Select products that appeal to people who want to buy something - even the fact that you can drive many people to your site, the reality is that the majority will not return that's the reason you should make sure you get the best conversion rate possible in order to make your efforts worth it. The most effective way to accomplish this is to select products that trigger impulse purchase for those with a lot of money available.

Pick a product that customers are actively searching for. You'll have a difficult time driving people to your site if customers aren't searching for it on the internet at all. It is important to utilize Google Keyword Planner and Trends to determine the popular phrases that people are using with regard to your specific niche. If nobody is searching for the product you're selling No effort you put into it will make sales, and you've failed essentially before you began.

If possible, design your own brand. People will be more impressed with your dropshipping company if you have the ability to brand the products you sell. If you can, look for niche products that you could white label and incorporate your customized packaging, branding and packaging. This obviously doesn't apply to all niche items however, it's recommended to apply this to at least one product you provide.

Offer products that consumers cannot locate locally. If you're selling the same items that the retail stores across the street, then all of your efforts will go wasted. Why should people patiently wait for your delivery of an item when they can easily purchase it at the local retailer? It is best to conduct your own investigation to determine what customers are looking for that they cannot purchase locally.

Examine your competitors - Is there many competitors in your particular field? Are you able to be ahead of your competitors? Are you willing to even consider competing? These are the kinds of questions you must consider when you are looking into your market. Most likely, there are already many participants in your particular area however that doesn't mean you can't or should you not join the party. You simply have to be aware of your competitors and be prepared to react in a manner that is appropriate.

Determine the challenges that your niche product can solve. In order to make sure that your dropshipping company will earn some money it is essential to conduct some study on the issues that your intended market is facing, and then find solutions to the issues. A great online resource can be used to conduct your research is Quora forums. Look for the kinds of questions that users frequently ask in your field and then think of ways you can find solutions to these questions using your product.

Step-by-step guide to choosing an area of interest

If you're stuck on the best area for your dropshipping business Here's a step-by-step guide you can follow. You can alter a few of the steps to suit your needs If you're interested.

Brainstorm about an Idea

This is the first step in any company, the ideation phase. Find hobbies or other interest areas that you are familiar with, and discover a variety of merchandise and products.

Take a look at your customers' demands. When you're thinking about your ideas consider the last product you purchased, and whether it met your requirements. It is easy to figure out what the requirements of customers are by analyzing them on the products you have purchased in recent times. Make a list of all niches that you imagine that meet the customer's needs and then determine the value they have available.

Get Your Thoughts Organized

It is essential to be prepared with a pen and paper at hand during your brainstorming sessions, as you shouldn't be able to keep all of the ideas you had in mind earlier. If you can, record your ideas on an Excel spreadsheet to be more well-organized. Label your ideas that are specific to their source and the specific motives for choosing them. Always take notes as often as you can, as it will allow you to revisit your thoughts as often you'd like.

Assess Your Niche Throughout the course of a year

If you're determined to start a dropshipping businessand are determined to make it a success it is not a good idea to jump into it blindly. If you decide to choose an area of interest, you have to make the effort to research it thoroughly prior to you decide to commit. One thing that you must be prepared for when choosing a niche is that they will have fluctuations and ups and downs through the year. For instance when it's Christmas time products related to the holiday season are selling out like hotcakes, however they may not sell as often, if all, throughout the rest throughout the entire year. That's why when choosing your niche, you have to establish a period when you'll advertise certain items, and then change them to other products that are sold during the same time period during the entire year.

Peak Season for Sales

According to experts in the field that online sales tend to increase in the Holiday season, which begins around the end of October. It reaches the peak around November, and then decreases toward the closing of December. The final three months are the time when online retailers are the busiest. It is possible to use tools such as Google Trends to guesstimate which products will be the most sought-after during November and December. Then, include those items in your niche list to start. It is recommended to begin making plans for your niche and how you'll market your site in September, so that your site can run full-time during the peak months.

Off-peak Seasons

When the peak season is over and you'll see your sales are decreasing But don't worry because this is the ideal moment to unleash your imagination. Research consumer habits in each season. It is important to determine the items that are most popular throughout the year. Find out what products fall into your area of expertise, and modify your listings for your products accordingly.

If you have an idea of a specific niche then you are ready to move on to the next stage of the process that is to search for potential suppliers. Remember that you are able to alter your focus at any time you want to and, in actual fact, you could try two areas of interest if you like and you don't have to a single niche.

Chapter 5: Niche Evaluation

Dropshipping Niche Evaluation Tactics

Note: In the free Shopify guide I offer tools that are extremely exact in selecting a lucrative area, the purpose in this section is to help you to make all decisions by yourself and distinguish the positive from the negative.

Products and niches can go hand-in-hand. Selecting a niche is an extremely difficult task, however, when you've executed it properly and you've done it right, it will be worth it. Many dropshippers begin by choosing products that appear appealing that they will "sell themselves" only to discover that their niche is popular and saturated with products, and they did not get much attention.

How can you make sure you do not pick one that is over-saturated in the face of millions of niches, with millions of products available to select from? What should you look for and where to be looking for?

The first thing to consider is when we start looking at niches, we must consider the competition level and the market demand for items and if the niche is a hit and has been through the years. When searching for an area of interest, consider the product as the primary focus instead of non-physical items.

However, let's examine how this works in real life.

Method #1 Method #1 - Search Engines

It's as simple to get started. Conduct a Google search for 'profitable dropshipping niches'. It's not difficult and only a few steps and you'll be able to spark your thoughts.

Look at the results of a Google search results.

Check out the items you've found and take an overview to help you get an idea on what's going on in the overall market.

They are usually being discussed and are niches and products that are being marketed to be popular. However, it's so easy to do that it's easy to become overcrowded.

However, one should not assume that that a niche isn't in terms of competition, not overly popular and the market isn't too popular, it's an excellent thing. A niche may be low in competition due to it having an untapped market and the need for products isn't high. That means that you have the lowest chance of establishing a successful dropshipping company based on niche.

So , how can you tell whether a particular niche is profitable? There are some requirements which must be fulfilled. It is likely that you have a particular area in your mind that will allow you to prove the answers to these questions:

There are enough physical products to dropship Dropshipping is built on physical items. If the area you're thinking of does not have many physical products or innovative inventions constantly on the market however, it is more focused on intangibles like software and digital products, then it is not the right market for dropshipping.

If you're thinking "doesn't all the products you sell mean the same thing?' then it's true. But, often times, certain items are not advertised correctly, and still be a great source of profit. However, if you look beyond the popular niches, you will find niches with enough products that aren't too in the market.

Also, search for areas that offer products with an appeal to a wide range of people. For example, a tailor designed item could be a little appealing to a limited number of people. But, a product that treats back pain will be highly niche specific and appeal to a larger group of people in this niche.

*Is it a fashion?

Contrary to what many believe they are not necessarily negative. However, one of the biggest mistakes made in dropshipping is to market a product within an area that is the result of an occasional advertising campaign. Consider the popularity of'slime'. Many dropshippers jumped onto this idea because they could see the speed at which it sold and the possibility of making an easy buck, which many dropshippers did.

The problem was that after the craze had exploded, and when people were making their own slimes, some dropshippers had any idea of what they would have to follow the success. Dropshippers who have this "fad-like" mindset constantly search for the latest trend and usually don't see the next one or spend money on products that are over-saturated.

In the absence of a fad There is no any real plan, strategy, or long-term plan for their company, just occasionally, short-term results. This is the reality for the majority of dropshippers in the present.

If you've come across an item like this, and I'm sure there will be many products that be like this, and that's why dropshipping won't be changing in the near future, by all means, join in on it, particularly in the early stages.

But it's not going provide you with a continuous flow of income. Always have a plan and a angle that allows you to change the one you're currently in. Better yet, choose an area of interest that has stood the tests of time. This isn't a bad thing that you can earn quick cash in a short amount of time. However, some people don't know what to do to move into other areas. All it boils down to is what you're looking for whether you want a short or long-term?

Is it an ongoing trend?

The fashions are however popular in a slow manner and are in high demand for a period of time. However, they don't last for long however, they are in fashion for about five years. This is better for a business that is sustainable and provides you with the chance to expand into new profitable areas when you keep up-to-date with what's happening.

Take note that with trends and Fads, this does not necessarily mean that they are better than fads to make money. However If you launch ads at the right moment, typically at the time of the initial hype, you can make lots of money over a short time.

Although trends and fads might not seem to be so important in dropshipping since you don't maintain or manage inventory or purchase stocks in advance it isn't easy to identify other trends and keep the long-term viability of your company.

Is this a market that is growing?

Some trends begin from the beginning and then develop into a steady market. When considering areas of interest, you need to consider if they really add value and is something that consumers will be using in the long term. If you are able to spot these aspects in the beginning, it's an obvious sign of a market that is growing.

Wireless drones, drones, and equipment for blogging video as well as drones are examples that are gaining popularity in a market. Why? They didn't exist prior to when they were created the invention of computers, they started to slowly expand and evolve into items that specific niches cannot do without.

Does it provide value?

Certain consumers who know about dropshipping, have given it the wrong impression often in good faith since some dropshippers set up an online store, include items, (which could be actually great products) put some snarky advertisements in prospective buyers' faces , and then expect it to suffice.

Customers can easily see right through it. There was no serious effort into what you did and had no clarity of vision. Your items will appear cheap, even though they're top quality specimens in the process of being discovered. You're basically not offering the items any worth.

If, however, you approach dropshipping as an actual business and invest enough effort and time into it, and provide products that have an inflated value and the appearance of being unique and it's difficult to determine if your shop is a dropshipping business that is not concentrate on your customers to the fullest extent, then you're doing this in the correct manner.

If you're unable to achieve this within the niche since all of these services are available and sold, then this isn't a niche for you.

*Are you too passionate?

If you're obsessed with, say cellphones, accessories and cell phones and think you're able to enter that market and make money but you're not, then you're likely to be wrong. There are a lot of physical and physical stores that sell accessories. If there is something you are passionate about few people are attracted to there will be lower competition, but you'll also experience a lower demand.

However when you find an area of passion within an area that is popular and in a position to create a unique product that no one is aware about, then you've discovered a gold mine. When choosing a topic focus on your own interests, think about what other people are passionate about, and which you know is highly sought-after.

Method #2 Method #2 AliExpress

AliExpress As mentioned in Chapter 2 it will be utilized extensively by dropshipping novices and even more experienced dropshippers, it's easy to use, reliable and is in use since.

Then head over to AliExpress and explore the various categories. Within the categories, you'll be in a position to view the items and consider whether the items are unique and if you can add value to the market by utilizing the products you've seen and so on. Take this product, which is a posture corrector.

It can help ease back pain, helps in improving your posture, and posture, and so on. It's also a distinct kind of product. I'm hoping you will be able to discern the niches that this product is a part of in terms of people who will be interested in the product, and who you could sell this product to successfully.

Method 3 Google Trends

I then went the Google Trends and typed in 'back pain' into the bar for searching. Here's what it returned.

The image above illustrates that in the past five years back pain has been of great importance. In addition you can look up the corrector of your posture at local stores such as Walmart to check whether it's available. To save effort, this isn't therefore, the product is distinctive. Most of your queries were already answered and you've clearly established your area of expertise and may even have discovered a product using this research technique.

Construction workers are also attracted by this type of posture corrector and it's possible to come up with a variety of reasons. That's another market right there. Let's look at this. You could think of other items that construction workers might be interested in, and you could give them.

It could also be a part of the gaming industry for those who are glued to their games or even playing games to earn a living. There are additional products that you could offer in the gaming industry in the event that you decide to go that route.

You just need to follow the steps I have explained. There isn't any difficult, costly method to figure out if you'll be successful and there's no need to be overwhelmed by this even though you could be able to spend some time identifying the best niche for you.

It is important to narrow your market, like I did by asking yourself the most fundamental questions and brainstorming ways to add value and ways to develop your business around this particular area.

Does Niche Always Matter?

No! Niche and products go in. You could be in a highly competitive niche however, you can offer unique products that have a high-value which other dropshippers don't offer and thus making lots of money by in the process.

In our case, we will look at the position corrector, it's definitely within the health category, which does not require us to do a lot of research to find it crowded with products, similar to the accessories for cell phones. In this instance it is possible to avoid any over-analysis of the niche since the product is able to solve an issue that a lot of people suffer from and, therefore, it is able to generate value. You can market these advantages.

It is well-known that there is a huge market in the health sector due to the abundance of products that fall into this category and , consequently, it's an industry that is profitable. If it weren't successful, the range of products would be restricted. The way you market your product and how you present your product people will result in a lot of results.

Chapter 6: Selling Things Online How To Find Items To Dropship

This is a vital aspect of this kind of business. Even when you know the area you're interested in but you might not know precisely what products to offer. There may be hundreds (and often even thousands) of items to pick from in a distinct sub-segment.

What you're looking for aren't just items to offer, but items that can be sold. There may not be enough inventory that you can dropship however, you're trying to reduce the amount of the space you have on your website for listing instead.

Another time-tested trade-secret (or rule of thumb) you should adhere to when choosing the right product: you must let your audience determine the final decision. It is your goal to sell to them and satisfy the need they have. This is the reason they determine the products you'll list on your online store.

One of the things you must do is seek out customer feedback and trends Google searches, as well as online chatter. The aim is to discover items that are popular with customers.

The characteristics of products that are sold More efficiently through Dropshipping

Can you sell any item in the dropshipping industry? Yes. It is possible to sell anything that you find on the internet. You can offer toothpicks for sale online If you wish.

But, it can't be said that there are some products that are more popular in dropshipping businesses, but there are also products which won't. Here are some features of the products you must look for:

The retail price for the product must be between $15 and $200 (this is also known as the sweet spot in the prices of items sold through e-commerce)

Items that are sold throughout the year, but seasonal products aren't as popular for dropshipping.

Any item which weighs less than four pounds would be perfect for this type of business model (consider the cost of shipping)

The products that are roughly what the dimensions of shoeboxes, or products that can fit into shoeboxes sell very well (i.e. any item that is light and small)

It is easy for products to market, meaning they don't belong to an industry that is controlled by big brands.

Unsaturated, and with minimal to no Brand Presence

Be aware that you won't be able to capture some of the market share if you're small-sized business if you're associated with a brand that dominates and is a dominant force in the market. For instance, don't assume to beat Apple or Samsung with dropshipping an alternative brand of smartphone.

International ePacket

It is also called"the" ePacket limit--4.4 pounds. The limit is set to prevent the cost of shipping your goods.

Be aware that the ePacket is subject to sizes that are both maximum and minimum. Refer to the following table:

Minimal Requirements and Maximum Limits

Length: 14cm (for containers) The thickness of the box is 90cm (for boxes)

Width: 9cm (for containers) Permissible difference (boxes) 2 millimeters

Permissible deviation: 2 mm (for containers) Maximum length (box) 60cm (permissible variation of two millimeters)

Length + diameter 2 = 17cm (minimum size for rolling) Length + Diameter x 2 = 104cm (max dimension for roll)

(for rolls): 10 cm Minimum length (for rolls) 10cm. The maximum length (for rolls) 90cm (with the possibility of a deviation of 2 millimeters)

Seasonal Products

These products are ideal for those who already have an established e-commerce store. But, if you're just beginning out, seasonal items could reduce your store's selling potential when the items are not in the season.

Potential Margin of Profit

It is important to take into consideration your profit margin. Anything that is sold for under $15 are found to decrease your margin of profit even when you sell the product in large quantities. However, any item with a price of more than $200 is typically difficult to sell, especially in the case of volumes of sales.

Of of course, there are some products that may have exceptions to these guidelines. There aren't lots of those.

Good Quality Products from Top Suppliers

We'll just mention this briefly to emphasize how significant an aspect this is in choosing the product.

Although a product may appear like it could be a worthy item to offer (it is well-known, addresses a particular need and there is a massive market for it) but if you're unable to find a trustworthy manufacturer for the product, it's probably not the best choice for you at the very least.

Potential for creating repeat customers

A good product must be able to bring the possibility of repeat business. This is one aspect that is often ignored.

This simply means that a product that is good is one that's capable of making a previous customer return to your shop and purchase again. This repeat business could be in the form of an item or component which requires renewal, or even when the product is no longer in use (i.e. it's consumable).

Perhaps, for instance, you came across an Vitamin D3 supplement that sells well and has little competition. It could be a great product because it could lead to repeated purchases from your existing customers.

After they've taken all of the vitamin D3 capsules that come in a box, they'll need to purchase another package or box. This is an example for the concept of a repeat business.

Another option is a pocket-sized photo printer. Yes, you had to sell the printer once (and it'll last for a long time). However, it's going to have to be ink-based and require a particular dimension of photographic paper which you'll also supply. Ink and a photo paper are designed to create the possibility of repeat business.

Fewer Breakable Parts

One of the common issues dropshippers face is botched deliveries that result in the product being damaged on delivery.

You could claim that delivery of items is not under your control. It is. If you make sure that you purchase an item that is made of less parts, and fewer fragile parts to boot--you are on an effective way of reducing the chance of customer returns and negative feedback.

Pick products that are strong with a limited number of removable or peripheral components. It is not possible to rely on the delivery service's ability to ensure that your product will reach its destination in the same piece.

Another thing you can try. Choose a provider who is known for providing high-quality packaging. A few good third party suppliers can do more than wrap the item using bubble wrap.

Additional Tips on Choosing products for Dropshipping

Examine all the products within your area of expertise

Cross refer to the possible products you've found with the top-selling items (do the comparison)

The products should satisfy the actual needs of customers

Beware of Pitfalls and Mistakes

Don't put too much pressure on yourself if you do make mistakes, especially when choosing an item to launch. In fact, you might even be over the sketchboard from time to time.

We'll make mistakes as we are in this industry, but it's okay. Our mistakes will be our teachers, a brutal and cruel teacher, at that. They will eventually aid us in our efforts to get better.

For you to avoid costly mistakes The best way to prevent making major mistakes is to take a lesson from mistakes other marketers have made. The pioneers are there not only to show you where you need to go, but also to demonstrate the traps you must avoid.

Here are a few of my biggest mistakes you need to be wary of:

The best choice is to select a product which is extremely competitive

This, I'd like to be able to say is a typical rookie mistake. I believe that everyone who dropsships will fall victim to this error at least once in their life.

What happened to me? There was a moment when I thought I'd found an amazing product, Bluetooth speakers.

I thought it was a wonderful product to market on the internet. In addition, I knew many things about the speakers. I had a few of those speakers for myself.

I scoured the market and saw that there were a lot of sellers selling the products on eBay and also on Amazon.

I look up search terms and see that it's an extremely popular product. There have been a lot of queries for Bluetooth speakers over the last months, and the trend shows that there's a constant and consistent demand for the product.

In addition, there are a lot of people searching for it through social networks. I noticed users posting these items for sale on Facebook Marketplace and a lot of inquiries have been made regarding it in the past.

I was of the opinion that it was a great product which made me think that I might take a bite of the pie. There are plenty of other retailers already earning profits from this and have already snatched up a significant portion part of this market.

I was thinking that there might be some hundred people that are seeking a different path. This is the place I imagined I could place myself.

Was it the right decision? It wasn't.

I've since realized the hard way that trying to squeeze myself into a very competitive market is a mistake that's waiting to occur. There is the possibility that I could capture a little percentage of market share.

However, I didn't expect an ongoing price battle that would be fought against competitors. I was just starting out and they had a long history. They are able to continue even in the face of low-performing returns However, I needed money to sustain a growing company.

The story is short, I ended in folding due to the pressure.

Selling knock-offs

I'm not suggesting that all Chinese manufactured products are imitations. The product doesn't need to be produced in China or any other nation in the Pacific. Knock-offs can be made anyplace.

The problem is that you'll soon find sellers of knock-offs and imitations. However, the FBI hasn't yet caught these suppliers as of yet (or another agency which is in charge of the catching of them).

Don't get me wrong. There exists an industry for counterfeit goods. The people who buy it. There's no doubt about it.

Here's some advice from my personal experience: avoid imitations. I haven't had to deal with it, however, based on what I learned from a colleague, it is possible to get in serious legal issues when you sell these items on the internet.

That's why you need to examine your suppliers with care, particularly when they're brand new or claim to offer the same items at a very low cost.

If their price seems too appealing to be real, then bet your money on it is. It is likely that they are selling fake goods. It is possible to be able to get away with it and create another website for their supplier. But if you're an unexperienced businessperson and you've just launched a brand sole proprietorship, then the legal consequences are stacked against you.

The most important thing to remember is that you shouldn't ever try it.

The marketing of a designer product

Designer items include all kinds of brand name that is available. They can also be extremely attractive to sell as customers are always looking for these items.

In addition, they are very expensive, and might lead you to believe that they let you increase the value of your markup. A final piece of advice: don't attempt to sell them.

It's also from personal experience. Do you think you can improve your margins with designer items? Let's face it - the profit margins on these items are extremely low.

They're a good fit for large stores and brands because they are able to live with low return on investment. They have pockets deep that they are able to hold onto. Not you, not the dropshipper.

The first thing to note is that your purchasing power is at risk. What happens if there are returns? Retailers with large stores could make that happen because they have more funds than us smaller retail stores. If you're not able to do that, you shouldn't even consider these types of products particularly when you're starting out.

The Best Dropshipping Products We've Found

I've saved you the effort of searching for the hottest products of the moment.

You can also look these products up using Google Trends and Keyword Planner to determine if they're still viable to sell via dropshipping.

Here's the complete list:

Kids tent

Artificial flowers

Pens for calligraphy

Tote bags

Ukulele

Sports bra

Matcha

Insulated bottles

Bags that are waterproof

Print socks

Cream for anti-aging

Seamless underwear

Organic tea

Smoothie blender

Baby Carrier

Bands of resistance

Kits to whiten teeth

Muslin blankets

Wooden watches

Smartwatches

Chapter 7: The How To Run Your Dropshipping Business

You're Running Your Dropshipping Business? Dropshipping Business

Then, you'll know the basics of dropshipping well-understood enough to think about launching a company that deals in dropshipping.

But, before you begin, you'll need to research the most important financial and business-related steps if you're determined to get your business public and the reason you are reading this book indicates that you're serious about dropshipping.

Certain of these steps should be done, and others are just beneficial to complete. But, recognizing the issues and addressing them today will help you avoid time and the challenges that could occur in the near future.

The commitment required

Like any other business developing a profitable business in dropshipping requires dedication and a long-term outlook. Be careful not to risk a failure too quickly, or too early.

You can earn a profit within your business, before you can achieve that you must take your business's strategy in a more realistic outlook and set realistic expectations with regards to profitability and investment. When droppingshipping is involved your primary investments are time and cash.

Time to Invest

The idea of investing your time is a popular choice when compared to investing money immediately. This approach is perfect because of these reasons:

Spending time on your business will allow you to better understand the ways that dropshipping works both inside and outside. The books you read can only take you so far. It can help you get started however it's crucial to get to know the basics by yourself, so you can help your business grow and expand.

Spending time on your business will help you to understand and understand your target customers and your market that will allow you to make educated decisions as your company grows.

By investing your time, that you're less likely spend a lot of money on "nice-to-have" initiatives that aren't a essential element to your success as a business.

The time you invest in your work also helps you to learn new skills that can make you more successful when your business grows.

If you want to leave your 9-5 job and focus every moment working on your dropshipping business, you and most people simply cannot afford it unless there is an investor willing to support you or have enough money to keep you afloat for six months at a minimum.

However, it's not impossible. You'll need to set an amount of time to work through the challenges however, it is feasible to continue droppingshipping even if you are working a 9-5 If you set the appropriate expectations for your company.

As you expand as you grow, you'll be able to slowly move to working full-time in your business, as your cash flow and profitability will eventually allow it and you will also have to concentrate more on your retail business, and this may affect your day-to-day work.

If you are offered the chance to run your business full-time, you should to make the most of the opportunity. This is the most effective way to boost your odds of success and boost your profits. Whatever you have time to spare, make sure you are focusing to marketing, especially in the beginning in order to establish momentum. Concentrating all your time on your dropshipping company and an intense focus on promotion and advertising your company will allow you to achieve that full-time salary of $50,000.

In doing so there are two aspects to keep in mind:

Once your dropshipping company is operational the tough part is in the maintenance of your site however, this could be less time-consuming than earning the same amount of money as an office job. Your most significant investment will yield by scalability and effectiveness that dropshipping can provide.

You'll be creating more than an income stream while you build your business. Additionally, you are building yourself an asset that could be later sold off. You should consider how much equity in your business that you're accumulating along with the cash flow generated when you evaluate your actual return.

Making Money Invest

It is possible to establish and start an online dropshipping company by investing a lot of money, however this isn't the best option. The greatest success you can get is when you've completed the work of your company. It is essential to be someone who is deeply committed to the success of your company beginning from scratch or have an individual who has that kind of commitment. It is vital both you and the business associate you have to know the way your company operates at every level.

Although investing large amounts of money into your dropshipping business isn't something you should do, you must have a bit of money, around $1,000 to launch and operating costs for hosting charges.

Determining a Business Structure

A key part of being serious about your business is to create an official company entity. Here are the most commonly used corporate structures you could explore for e-commerce:

Sole Proprietorship

This is a basic and easy business structure you can use in the dropshipping industry. But, it also means that it gives no protection against personal liability. This means that should your business is sued the personal assets of your business will also be a part of the calculation.

However the requirements for filing taxes are not too strict and all you have to do is ensure that you file your personal tax returns as well as your earnings of your company.

Limited Liability Company (LLC)

An LLC can ensure that the personal belongings of your clients are secure since you can form your company as an independent legal entity. This type of structure offers greater protection than a sole proprietorship, but it's not foolproof. For this type of structure, you'll need to follow certain filing requirements aswell in paying fees.

C Corporation

C Corporations, which when properly set up, provide the highest level of liability protection. majority of companies are formed with the designation of C Corporation. This type of business structure is costly and also is subject to double taxation since the profits do not pass directly to shareholders.

Which structure do you prefer?

A majority of business owners opt to LLC or sole proprietorship. Dropshipping is advised to choose an LLC since it gives more independence from personal finances, cost as well as insurance against liability.

Ensure that your finances are in order

If you are starting your own business, you should not be a fool by mixing your company finances with personal financial situation. This can cause confusion and be an the perfect way to create a nightmare of auditing on the professional and personal level and makes accounting more complicated also.

Separating your personal and business finances in a separate account is the most beneficial tips that you can follow regardless of the type of business you're in dropshipping, brick-and-mortar or dropshipping. Setting up new accounts under your company's registered name is an excellent method to begin.

Here are the banks that you should get open:

Business Checking Account

Your business's finances have to be consolidated with one primary checking account. Every business's revenue has to be deposited into that account and all expenses related to your company must be taken from it. This will make life easier for the business owner and your accountant.

PayPal Account

PayPal accounts are wonderful to include in your online store, especially when you intend to accept payments through PayPal. In this case, you'll need an account for business that is linked to your website's e-commerce.

Credit Card

Because credit cards are a preferred method of payments for providers, they are the preferred payment method also by your clients. Similar to that it is recommended that you be using a credit card for your business to cover your costs of running your business as well as for inventory purchases to dropshipping.

There will be a lot of purchases you make at your supplier and using a credit card linked to your business account allows you to collect some substantial rewards. Check out cards that reward you for travel online, transactions on the internet and other kinds of products which you're likely to be buying more often.

Collecting Sales Tax

Taxes are something you should be aware of. You'll need to pay the sales tax when you are in the following category:

The state that your company is operating in has sales tax.

An order is issued by a person who is a resident of the state you reside in.

If you are a resident in other states or in states where you have their own sales taxes, don't have to collect tax. The tax laws applicable to online sellers are advantageous, especially when you're new to dropshipping and are considered to be small.

Local Business Licenses

Companies require business licenses, and these licenses must be renewed regularly. Dropshipping companies, the regulations may differ as well, and they may involve operations out of homes offices. If you are considering starting a dropshipping business check out what law and regulation in your area demand and if there is anything.

Chapter 8: How To Plan Your Dropshipping Company

If you are planning your strategy it is essential to be aware of your place on the internet and how it will affect your company in various ways. For instance, you're likely to be hoping that you will get clients via the internet search engine. It is impossible to be optimistic in this scenario, but it will likely not pay for the expenses. Finding a site that is well-positioned in the web search engine is a matter of skill, money and time.

Many people have the information, but there's a certain art to be practiced by a website to achieve an outstanding internet search engine ranking. If you're not one of the few who have this talent in your genes, you should not neglect the search engines and then, you can be that you're an SEO professional.

In general, you should expect to finish at the top of the food chain and the customers you'll receive will be leftovers from the major websites. They'll appear like a person looking for something the major websites could not provide. Thus, you must conduct your own research and try to find out what these leftovers from the major websites are looking for, what they require and the best way to satisfy their needs and desires. There will be people who are on the rebound when they look on the top websites for pricing and information.

You must be prepared to capture the swoopers as they run and then stop them on your website. What's the highest amount you can Earn? DropShipping, also known as DropDelivery, has helped to create Internet millionaires. If you plan to implement the particular method and become millionaire.

However, you should begin by establishing the general idea that you'll be taking on major merchants and large shops are found in malls. In order to compete effectively, you need to become familiar with the market. One thing to remember is that your prices will continue to increase, making your margins for profit will be lower. To make it easier for customers to discover them, you have to provide a price that can make a buyer want to pay in advance and perhaps spend more than what they buy at the local store or on ebay.com or any other online website. It is essential to be a value reseller who is able to advertise its business in as many ways as possible. People are on the Net to find information. If you can provide them with sufficient information, product reviews and other kinds of information, you be able to add value. If you can provide valuable information and push your site's content with a lot of force then you will succeed.

As a unique persona and brand, you shouldn't try to be a competitor. Join them and sell to them. A highly efficient drop shippers makes use of Amazon.com along with various affiliate networks. The risks they face are site configuration, promotion costs as well as time. If you're not familiar about this program Amazon Affiliate Program and what it can offer you, look here. Note that I make none of my earnings, either in any way, directly or indirectly by advertising Amazon. There are many people looking to retire So, try to learn about them and precisely what they've made it happen to establish an organization for dropshipping.

Beginning A Drop Delivery Organization.

If you're planning to establish a dropshipping business There are a few things to take into consideration. When you first start the Drop delivery service, customers concentrate on the product and then spend the time searching at a drop shipping service that can provide the product they want. There's much more to starting dropping delivery services that should last for the first year. You could have the top product on the market but you'll need an effective sales process, a marketing plan and an SalesForce System and customers. To attract customers, you must develop a comprehensive marketing strategy that you follow. Consider your place in the declining the sales of transportation. It is important to understand that you're important to the suppliers. In order for them to be successful they require salespeople and that's exactly what you're doing. Your company is an independently-owned service. You are a commission-based agent. They do not spend money on your behalf. You build an online site, market it, and spend an enormous amount of money and time finding customers for their products. You work hard and invest in the promotion and then deliver the product. You're one of thousands of people looking for products to sell to Drop ship distribution. As a drop shipping sales representative, you are able to assist these factories in selling their products retail or wholesale and the dropship provider typically competes with you by selling similar products at a lower cost through their site. You are valuable and also a liability and thousands of people are for you to be replaced.

Some earn quite a bit of money. the methods you earn money often prompts concerns from those who want to know how much they earn. The answer is easy that you purchase at one price, and then sell it for an additional charge. The difference between the purchase cost and selling price is the gross profits. In order to calculate the revenue you earn from your website you must subtract your costs. If you sell on behalf of a company such as Amazon that is a major retailer, you will receive a portion of every sale. If you sell something similar to Polish pottery you purchase from a factory or supplier, you calculate your gross profits by setting the price of the item. The money you earn is from buying cheap prices and charging the lowest price it is feasible to keep buyers returning to buy more. What you have to do is conduct a thorough market research, locate products that match your preferences and then look at the price and then determine the best price you are able to offer the product.

To determine if your business is profitable and not must to figure out your expenses and the value that you put into your work. If you do not understand the cost-based framework, there's no way to determine your profit. Once you have resolved the "Service Plan Industrial Zone" The number came as being very clear.

Who Are The Drop-shipping Vendors?

Not all factories ' suppliers and merchants are a part of drop delivery. They don't want to be associated with withholding the delivery of individual items and supplies because of the same reasons they don't want to be. There is generally lots of tracking of deliveries, and there's an enormous amount of work involved in keeping inventory and also in the shipping of individual items. Therefore, some people can fill in the gaps by offering supply and fulfilment solutions. Most manufacturing facilities don't have any supplies at all; they just produce according to demand. This is quite common for instance in the crystal and glass business , where manufacturers ask for orders that range from hundreds to thousands for any particular product. They do not have products to sell.

Thus, the gap is being filled by business owners who buy large quantities of items they store in their warehouses and offer other satisfaction solutions. They are commonly referred to as wholesale decline sellers. They are vendors who stay in business to make profits, so the price that you pay for goods in this context will be greater than what you would be charged if purchasing directly from the factory. Sometimes, the price could be the rate at which manufacturing facilities operate. In the extreme the drop shipping cost could be higher than the price of the product purchased from a major retailer. Large sellers will pose a lot of opposition to you in both offline and online. They realize that you will not make money selling the product. Earn money by buying items at the lowest cost possible and selling the item at a lower cost as is. They rely on a the highest turnover in order to earn a profit.

As a small drop-shipper as a drop shipper, you're not able to negotiate good prices since you're purchasing only one container per time instead of having to deliver containers in a single batch. It is impossible to compete with them, and you'll be competing with the same major merchants and manufacturing facilities that are listed on auction sites like eBay. They usually sell for a loss and create mailing lists for customers and customers.

Dropship providers you will work with are those who purchase lesser quantities than the top merchants , and they offer prices that are usually higher than the prices you'd pay at an individual retailer for the same item. In addition, the prices they provide you with will be more expensive than direct rates from manufacturing facilities and the wholesale cost that is typical.

What Can You Offer?

I've heard of having a passion for your product. This is a path to failure. In this instance, people are enthralled by Polish pottery. They love it, and would like to market it. They get involved and realize that it takes more than enthusiasm to sell Pottery. They find that the market is crowded. They are not only competing against thousands of small sellers, they're also competing against wholesalers and manufacturing facilities creating mailing lists to and sell on ebay.com with very low costs.

The ones that stand out against other competitors, put their interests aside and concentrate on the planning. They define their area of expertise and operate in their own particular country. Your particular niche within the world of drop shipping delivery is to define your region in the shipping industry and define your field of operation. You should choose the right the products or methods to market those products that make you distinctive and beneficial to people who come across your website as well as your phone number or your advertisement any other marketing materials.

You must do something valuable to attract people to pay for a higher price to buy from you. You must offer them something special. They should be planning to visit your small area of the world - your area of expertise.

Definition of an "Niche"A particular niche could be described as an event or activity that is specifically related to an individual's needs, interests capabilities, skills, or interests and also to the specific location of the need for a product or service. product. One example of niche advertising is the marketing of a custom. Hand painted Christmas spheres for churches to be used for fundraising. The client is specified narrowly and the product is defined the product, the service is targeted and the item meets the need. A further example is the fact that they are going to be the authority in providing only the peacock-patterned Boleslawiec Polish Pottery as well as offering every product produced by the factory. Additionally, there is an online site that is a specialization in marketing elliptical exerciser training device. Take a look at how easy it is. It uses Amazon.com to drop ship.

The importance of identifying your niche Here are a list of issues which you need to note down and take the time to provide a thought-provoking answer to. If you respond to these questions must be honest as your response will determine whether you are employed or not.

When you have a list of answers to these questions If you are considering establishing an e-commerce store that is in conjunction with a large retailer like Wal-Mart House Depot, Target, or any other chain of stores with the kind of product you plan to sell. On the internet it is likely that you will be competing with other stores that are online match-ups of these stores, or those stores. It is essential to establish an identity and offer that draws customers away from the big stores to your tiny, untested business. What is the best way to stand out from the companies offering the goods you plan to sell? What is your distinct selling point? What distinguishes you from the rest of the pack to make people want to purchase from you? What are the reasons why buyers will choose you? Are you offering convenience? Price? Solution? A presence off-line? Are you providing what you intend for sale or do you know what customers would like? What are your customers' needs? What exactly do you know about this? Have you conducted market research? Have you completed surveys? Are you merely coming to a conclusion since you believe that people would be enthused to purchase from you? What are the reasons consumers should believe in you? What react to? Do they really need you to satisfy an expectation that they can't get elsewhere? What advantage can you provide? What will make customers buy from you again?

Random Words on Random Things. There are three components to any formula for business Cost, Quality, and Solution. A customer only gets three of them to support it. Three of them are invariably unsuitable. If, for instance, you're looking to get a lower price and quality, you will not get the service. If you're looking for the solution, you'll have to pay more. As a service provider you aren't able to offer the three options and remain with the company.

But, when the definition of your area of expertise, you must utilize a the price, product or service that distinguishes you from others and makes people desire to purchase from your. In this instance, it could be that you plan to sell Boleslawiec Gloss ceramic. There are many wholesalers as well as major retailers who sell this type of pottery online. They import the product by the container , and hold several hundred thousand dollars of inventory in their warehouses. They have a substantial budgets for marketing and advertising and are also popular all over the country. You are planning to market Polish pottery, but you can only be able to afford an investment of $500 and $1,000 in your inventory. Your costs for import the product from the factory can be as high as the cost of supplies you pay the factory.

The price of the item displayed on your shelves may be more expensive than the price some people purchase the same item on ebay.com or at the major Polish clay stores. What is the best way you are likely to fight these People? It is possible and everyday in a variety of places across the world. One way to accomplish this is to design something unique for the people of who live in your state, town or a specific area. You offer the service or provide a convenience that will make people pay more for help you rather than to purchase something on online Internet from the big companies. You must make your own, and also make your little business more valuable to customers. It is possible to do this by taking your small inventory and making it an online directory shop where people are able to visit your samples and order online from Polish ceramic catalogs you have in stock. You could offer an exclusive order service to local businesses and set orders so that you are able to make a sufficient order through the Polish manufacturing facilities in order to receive substantial discounts, which will reduce the cost of the stock in your store.

Here are some other Ideas on other Products. Perhaps you could call companies such as banks, insurance companies and beer distributors, suppliers of food retailers, and so on and persuade them to purchase custom designed Christmas ornaments that are made to their specifications and used as gifts to their clients and staff members. You might be able to start an organization that provides sellers items on ebay.com and at the flea market. They constantly try to find products to sell, but don't have the funds to place large orders. Group orders can result in the best price for these. You might consider selling ceramic mugs featuring business logos to promote items. Whatever you do or the way you do it you must be sure to take the time to analyze and determine exactly what you intend to provide that will make customers want to purchase the item, which many thousands other sellers are purchasing by you!. Define your particular niche. And after that: Niche has done, Niche and Niche further.

The importance of a Practical Strategy for a Company. Strategies are commonly thought to be sales tools to impress bankers and financiers. However, your primary challenge is to be able to endure. Your strategy should be a tool. It should be practical. It should be a blueprint. Additionally, you need use it as the builder uses plans to build the home. It should be in line with the rules you receive when you an item of filled furniture. It should be precise. Below is how you can begin. List all the expenses you'll incur. Calculate the profits on each product you sell. Estimate the amount of products you need to sell to pay for your expenses. After that, you should think about how you'll attract enough customers to pay for the costs. To accomplish this, create an elaborate online marketing strategy. Develop a comprehensive web-based site plan, develop an extensive list of email structure strategies, design an elaborate offline promotional plan, and create a thorough voucher marketing plan. Keep everything you've into an organized three-ring binder. Utilize separators to organize your information. Make them up-to-date as you become "smarter". Make use of them. Use the guidelines. Set your goals and develop your strategy. Take the time to study how to apply each one of these strategies and follow-up with extra time to develop your own plan. If, once you're finished, your business doesn't generate any income in theory and it doesn't earn cash online. If it is able to earn money, there is the possibility of earning. If your plan is successful. Therefore, you should create a great plan. Information on how to make the plans is available all over the internet.

Chapter 9: Rich Are In The Nichs: Choosing The Products You'll Sell

The aim for this article is to assist you decide on the niche you can sell your goods in. There are two methods you can pick from - either establishing websites that are in response to your passion, or creating a niche-specific website with a profit potential absolutely. Whatever method you decide to go with there's a procedure you can use to determine the niche that is the best one to begin your venture. It's not a good idea for you to put in many hours creating websites only to discover that you're unable to earn a profit from it. This chapter hopes to stop that heartache and pain the readers.

Before we go deep into the specifics ahead of time, we'll go over some definitions. The term "niche" refers to a type of category or subject that defines the products you'll be selling. For instance the pet niche could offer items that are that are related to pets. It is possible to narrow your niche by making it more specific. For instance in the pet industry If you wanted to sell "dog beds," your niche would be more narrow. To narrow the market for dog beds, you'd concentrate on selling "dog beds for pitbulls. The more narrow your target market is, the more easy it is to sell to customers in that segment. Therefore, the niche could be considered to be a macro-category. Niches are made up of a keyword. The keyword of the niche is what people seek when trying to purchase items in your preferred.

Here's a method to help you select your area of expertise and the products you will sell on your dropshipping shop. Get a piece or paper and note down five problems that people experience and five worries people face and five topics you are enthusiastic about. From this list, select the top 10 things you're interested in selling. If you require a more help look into the following topics:

Health-related - This field covers numerous subjects. It is possible to experiment and treat common illnesses or the most embarrassing ones such as hemorrhoids and STD and locate products on the subject that can assist people.

Teeth Whitening : This area is never-ending because people constantly desire to brighten their teeth. You can offer a variety of whitening products available at your shop and narrow your focus by determining if they are vegetarian or non-vegan, and so on.

Yoga - This field is a favorite among so many people, whether they do yoga for health or for spiritual motives. There are plenty of products you can sell within this field.

Drones – The interest in drones is only getting started. It is the best moment to start experimenting with this exciting field and make an internet site for the novice or fan. Drones can also be costly and come with a variety of accessories to allow you to make plenty of money in this field.

Guns - People are awed by guns. This is a booming market with plenty of potential earnings.

Dating is a particular niche that can aid lots of people. Since communication has been evolving constantly it is possible to offer products to individuals to enhance their relationship.

Decor - There's there no doubt the decorating segment is a wildly popular area with numerous offshoots you can pick from. With a variety of rooms you can embellish in a home, to different kinds of residences you can create, there's a lot of options within this field.

Weight Gain or Loss - Weight loss or weight gain is a condition that affects many people. This area is sure to bring you money, particularly if you can assist people in losing weight easily and there's any way to make money from this area.

Weddings - the wedding market regardless of whether you're targeting people who prefer a more affordable wedding or an extravagant one This is one area that you cannot go wrong.

Being Pregnant is another area to focus on. There are a variety of approaches to explore this area, including topics on sex methods to achieve the sexuality of your child you desire and IVF options. This is an interesting area certain.

Pet Niche Pet Niche - Who doesn't have a love for their pet, whether they're cats, dogs or other pets. This niche may concentrate on accessories for pets or the essentials for pets to treat common ailments. Pet owners are devoted to their animals and would love spending money for their pet too.

Survival is a booming niche due to the fact that the majority of people are planning for the end of the world, and becoming more self-sufficient. This is a niche that focuses on helping people prepare, whether it is making underground shelters, making the effort to grow their own food, or even making a homestead. This is a fascinating area that more and people are becoming more attracted to every day.

Beauty - There's no way to not earn money in this field. No matter if you're focusing on cosmetics, hair, skin or clothing, the beauty market is constantly awash with trends to discuss and plenty of the potential for earning money.

Nutrition It is a well-known subject because there are numerous diets and lifestyles one could write about. Nutrition is a very well-known topic and you could approach it through discussions of food and diets, kitchen appliances or even vitamins for examples however the possibilities are endless.

Baby Niche Who doesn't love their children and who doesn't enjoy spending money on them? The baby niche is a great one to approach by many different angles, including reviewing baby products, or writing on the most recent fashions. This is a win-win-win niche with no doubt.

There aren't all areas you can market your products in. This is simply an idea to move the wheels. After you have a list of topics you are interested in, it's time to evaluate whether these niches are profitable or not prior to deciding about your store.

Making use of Amazon is an excellent method to determine how lucrative an area of interest is. It is easy to visit Amazon.com and then browse categories. Amazon categories. Select a category that you like and then take a look at the sub-categories. Also, search for accessories which you could offer in this area. Amazon is an excellent way to find out whether there are any profit-making products for your website since everyone uses Amazon. In general, if your field has a market that consumers want to invest money in, you'll be able to earn profits from your specific niche.

After narrowing your niche idea down, you'll need to search Google to determine if others are looking for information that is related to your field of expertise. Utilize the incognito search feature to ensure that your browsing history won't affect the results. Consider 10 keywords that someone might type when looking to purchase something from your area of expertise. If you examine the results of a search that pop up, what websites appear at the top of your list? Do they appear as forums, or brand websites? Are they small or authority sites? If you own a large number of smaller websites, forums and web 2.0 articles such as Reddit and Squidoo. You might be able to rank on that website. Keep this list handy as we'll require it when we try in the coming up of new product ideas. Keep track of your findings to remain well-organized and focused until you've come up with an idea that is successful. Here's how to approach this procedure. When you're searching for something to purchase What terms do you type in the Google box? Are you searching for reviews, testimonials, or discounts and offers? Someone seeking out vacuums should not enter an exact phrase like an a branded vacuum cleaner unless they are certain this is what they are looking for. The search terms they might include could be 'the cheapest vacuum or'replacement bags to vacuums. These kinds of terms indicate intent to purchase. You should utilize search terms such as those to discover the results Google offers when you search the terms or subjects that indicate that people are planning to purchase.

After you've got your niche-related ideas and product ideas, the next step you must consider ways to promote the website. Are you able to develop an identity around the site? Can you create an image that represents the website? Are you able to sell the website to someone who is looking for a long-term deal? This can help to narrow your focus further. If you need help coming up with names for your website, you can use this resource: https://anadea.info/tools/online-business-name-generator. If you're thinking of names for your website, be sure not to use the primary search term you have stumbled across during your initial investigation. It is not advisable to make the name of your website the keyword phrase since Google could overlook your site. Find a an appropriate name that is related to your field, however you could also name it. For instance, if we take a look at the vacuum cleaner scenario it is not advisable to choose a specific vacuum cleaner as your website's name. Maybe VacuumBuyingGuide or VacuumPurchaseHelp is a better choice. Let your imagination guide you to come up with a name which will inform visitors who arrive on your website what your website's purpose is.

Keyword research is crucial as it allows you to know if people will find interesting in your site or not. Keywords are simply the term used by people when looking for something on Google. The website tries to be found by a keyword and, when someone type the word into Google and clicks on the link, the site comes up. When ranking for a specific keyword it is best to be the first page that is displayed on Google. If you are able to be the first to appear on Google it's a great thing. The process is selecting keywords to rank for that allows you to be on the first page of Google. Making the right choice of keywords can help more people access your website which can mean increased revenue for you over the long term. The most important thing is that effective keywords can be the difference between success and failure for your site, which is why it's crucial to select keywords that will help you sustain your website.

There are two types of keywords: long tail keywords and short keywords. Short tail keywords refer to something that is broad, such as "healthy diet" or "kitchen appliances. These are difficult to achieve a ranking for, but the term "long tail" is comprised of at minimum three words and is very specific to the specific problem that your targeted customers are trying to solve. For example, a lengthy tail keyword could be "best vacuum for hardwood flooring. It is much easier to rank for keywords with long tails because the people who enter the keywords are seeking solutions. It is crucial to consider this since you could be passionate about something but if the people who aren't looking for the subject, then it is unlikely to earn any money. This allows you to determine if your topic is able to generate revenue or not. You will need to utilize Google Keyword Planner for this step. Google Keyword Planner to aid in this step as well as to use the Google search box is itself. Some paid-for tools include Market Samurai and Long Tail Pro. For beginners, you should focus on understanding the process. Then, you'll know whether you want to invest in expensive equipment or choose not to. Let's now look at some of the keywords we used during our initial study to determine topics using the Google Keyword Planner tool.

It is a tool for free that you can use if you have an Gmail account. To use the tool, enter Google Keyword Planner and then login. It is completely free and you don't need to pay but Google might provide the impression that you need to purchase the campaign. Google will want you to purchase something, so the first page may seem a bit too aggressive, but there's an alternative to it. To avoid the pop-ups when you sign-in take these steps.

Once you have signed into your account, Google may ask a couple of questions regarding your primary goals in advertising, but you must you must click on the tiny letters in blue that read "Have you had a good experience with Google Ads Do you have experience with Google Ads?', so you're able to skip this step.

You'll be directed to a page asking regarding the type of campaign you want to run. Choose 'Create an account with no an account in blue' under the boxes.

You must then confirm your business details. This will bring your to the homepage.

In the upper right corner, you will find your Gmail account information, but go to the left side and click on Tools. The first tab on the left side will be labeled "Planning.'

Under "Planning" you can find the Keyword Planner. Select the Keyword Planner. You might need choose the 'Explore Account first. This allows your access to your Keyword Planner.

If you choose Keyword Planner, once you have selected the Keyword Planner option, you will be able to choose between one of two alternatives: "Find Keywords and get search volume and forecasts.

You'll need to choose the 'Find New Keywords Make sure you select the US is chosen as the primary country. This refers to the number of people who use that search term on a monthly basis. This will aid you in finding keywords which aren't as popular, but can still aid you in ranking on Google.

Input your search keywords. You can input more than one keyword at each. You should also look for keywords that use questions like how you can, what, why and so on. These keywords will be used in the descriptions of your products on your website.

Here are some additional search terms you can apply to identify keywords for your site.

Intent to Buy - Based on your keywords ideas to search for other keywords on the tool to suggest what you will be made. Keywords could be 'discount', coupons, reviews"best" or "quality. Another way to do this is to think about keywords. easiest method is to consider what keywords you'd enter before buying something , or when you are planning to purchase something.

Cost Per Click - If you type in keywords, a price pops on top of it. The price is the amount that advertisers pay Google for each time someone types in the keyword and clicks their ad. If the price is very high it means that individuals are willing to pay huge sums for the keyword, which indicates that it's an extremely lucrative market.

Trends - Another aspect to think about when choosing your subject is the current trends. Are you looking to market your products in a seasonal market that is only available on specific holidays or are you dealing with an evergreen market, which allows it to be sold throughout the year? It is beneficial to mix seasons and evergreen niches as you're more established however, when you are just beginning out, you might want to concentrate on niches that are evergreen.

It is the Monthly Search Volume - When doing research on keywords Try to locate keywords that have an average of between 800 and 5000 monthly searches. Try to rank for it. This number could differ for different individuals. Many people look for keywords that have more than 1000 search results per month, up to 20,000 search searches per month. As you gain experience the ranking of your websites you are able to choose this amount for yourself but beginning with 500-1000 monthly searches will assist you in ranking faster and with less effort than a number with a higher volume.

Another useful tool one of the best tools to use is Google Trends. It is possible to go to google.com/trends and search for your product's niche and ideas there. Google analyzes the keywords you are searching for by month as well as the volume of searches per month so that you can determine if users have an interest in what you offer or not.

Once you've got your keywords, make sure you save them as you'll utilize them when making your product listings on your website. This is known as your SEO on-site. SEO is a term used to describe SEO, which stands for search engine optimization. This is a way to ensure the keywords you use are placed in proper places so that to get you to the top of Google. It is not advisable to compose those keywords in a manner that doesn't sound natural. Try to keep it as natural as is possible or incorporate keywords into your writing in a manner that is logical. But, you must remember to place your keyword in 3 important locations. The first location to include your keyword is within your URL. The majority of websites let you modify the URLs of your pages therefore, you can add your keywords in the URL. The next spot you'd like the keyword to be included is in the meta-description. If you search for something on Google the information that appears under the site that appears is known as the meta-description. The last place you need to include your keyword is in the description of your product. It's easy to become entangled in SEO's rules but what you need to keep in mind is to maximize your website's performance. Utilizing keywords allows you to make use of Google to bring customers to your site and differentiate you from your competition. There are a myriad of websites that sell dog collars, but paying attention to what breeds of dogs and the reason why your collars differ from the descriptions of your products will allow you to sell more items. Avoid falling into the mistake of thinking that price reductions will help you compete. You could still make an impressive profit, and offer a distinct advantage that sets you apart from other companies. If you're conducting study, take a look at five competitors and take note of 5 aspects you're doing differently. This will help you stand apart and increases your chances of success greater. You could even achieve more success by focusing on the way your product will solve the problems of people who buy from you and not anyone else.

Being able to identify a niche will allow you to be more specific when it comes to marketing This will enable you to develop more cells. A lot of people choose 18-65 as their primary market, but this is far too broad. Try breaking that market down into segments and be as narrow as you can. Remember that the riches are found in niches. When you are deciding on the demographics that you want to target Also consider whether you will be working with the kind of people who buy your products. If you sell, you'll need to interact with customers at some moment. The process of narrowing down your market can be a lengthy process, however, it is crucial to conduct a thorough study to ensure that the market is profitable. It's not fun to put in all the effort in setting up your website for your specific niche, only realizing that there aren't any products that are profitable that you can offer. Therefore, make sure you do your research. Don't take this lightly. step. After narrowing down your list of products then it's time to start finding a dropshipper to sell your product.

Uncomfortable to contemplate

Every business model has pros as well as cons, and it is exactly the same for Dropshipping models. It doesn't mean dropshipping is a bad business model, but it has to take into consideration the drawbacks specific to dropshipping. Consider the pros and cons and avoid the ones you cannot and come up with solutions whenever you can. Being able to deal with the negatives does not mean that your company isn't a success It is possible to avoid numerous pitfalls by making backup plans and implementing solutions.

Margins are not as high

Many are frustrated by the lower margins of their particular field. Customers want higher returns and increase their profits quicker. If you're willing to begin small, stay focused and study how to run dropshipping stores in the highly competitive world of e-commerce, you'll increase your growth rate steadily and reap the advantages of your hard-earned effort.

Market competition

E-commerce is highly competitive and sellers have to be able to outsmart one another to earn profits faster. Dropshipping businesses are simple to start with a modest start-up capital, making it a highly successful business strategy. The best approach to combat this is to begin your business right, and provide top-quality services to your clients and make use of all the available resources to build an impressive reputation for your business.

Issues with inventory and conformity

In a dropshipping model, you are unable to manage your inventory and are not able to monitor the flow of stock. You rely on wholesalers and suppliers, who serve a variety of other retailers. This can cause issues for products that are out of stock when you make your purchase.

Your suppliers are accountable for conformity to the specifications of your product, and mistakes and delays can happen. You'll have to manage your angered customers, and it is advisable to create the most backup supply feasible to allow them to use alternatives to your products.

Online retail platforms

Online retail platforms can be an excellent benefit for dropshipping firms. However, they can be negative, since consumers are able to bypass you and purchase the same items directly from retailers.

Shipping can be complicated

Dropshipping companies typically acquire their items from a variety of wholesalers and suppliers. Customers can be dependent on numerous orders for items that originate from multiple suppliers. This means that you'll be charged distinct shipping charges for each component of the law governing the different items, and if you do not transfer the additional cost to your client, so you must pay the extra costs. This will reduce the profit margin.

Providers aren't immune to mistakes.

Suppliers make mistakes and if that happens they are held accountable. Your liability is on you because your customer placed the order through you, not your supplier. The supplier's mistakes can arise from different reasons, some are legal errors, some caused by issues in the company of the supplier.

It is important to carefully research the suppliers you use to reduce quality errors in art, substandard packaging materials, as well as damaged or missing shipments. If your suppliers disappoint you because of their incompetence, you must switch to a trusted provider as soon because these mistakes will be reflect in your. Your reputation as a business is extremely vital, and you don't wish to hurt it.

Provider Skill Scale

There are many providers that are proficient, and some might not be able to grow with your company as it expands. The good thing is that you're able to connect to suppliers from all over the world and online retailers to purchase your goods which makes it easy to switch suppliers.

Chapter 10: Where To Purchase The Product And Where

Receiving the ordered items of customers delivered to them is not an issue. However, before that you must look for the best dropship provider to run your dropshipping business. It is interesting to note that a variety of dropship companies are available. After having made the research on their products and their experiences in the field then you'll be better positioned to choose the best one for your company. You might need to collaborate to two or three vendors if you are working with a variety of dropship items. A lot of things have been discussed about dropship companies before. We don't need to re-examine the elements that drive the process of selecting a supplier. We're concerned with what a dropship retailer can do to aid customers purchase the dropship items. We already know that these items will be delivered by a dropship suppliers, who are as well the maker. In light of the above it is not dependent on the location where the product are delivered from. The question of "where" is settled. The only question is how products are purchased.

You already know that dropship sellers don't have warehouses or offices in which the goods are stored. They don't even own the items they're selling. They may not even be able to see the product in any way, whether at the stage of production or when they are shipped to customers. Dropship companies utilize their platforms to offer products through dropship to customers who placed orders on them. They serve as the bridge between customers and the dropship company which is also the manufacturer. Manufacturers are always striving to improve their products in order that consumers can gain more value from their products. As they go along they get too busy to have time to promote their company or promote their products. The result was a vacuum that was created as a result of this. In order to fill the gap dropship retailing came into existence. As it stands now clear that the job of a dropship merchant is crucial. Dropship retailers aid their customers with getting their products delivered to their doorstep and assist manufacturers sell their products.

Finding The Right Product

Search on the internet for the product you'd like to purchase. After that several websites which sell the item will appear which you can select one of these websites. Search engines like Bing, Google, and Yahoo! can be used to search for products you're planning to purchase. Check out the shops and websites which are displayed one after the next particularly if you wish to compare prices. Amazon is another site where to search for products you're interested in purchasing. While Amazon offers products however, it can also connect you to many sellers. A lot of dropship providers make use of Amazon to market their products and customers who purchase items through the platform can also make use of Amazon's Amazon payments system. Don't rush to order your items through Amazon. Why? Vendors can sell their used items on the marketplace as well. If you are planning to purchase new products, make sure you do an extensive research to ensure you don't end in buying second-hand goods. Other than searching engines as well as Amazon as well, you may also purchase items on auction websites. Auction sites offer products that are much less expensive than those sold by stores. However, securing potentially lucrative bargains or unique items from auction sites can require more time. Additionally, auction sites are subject to many rules and rules. Do not bid on items there If you're not acquainted the rules and regulations of auction sites.

Alongside big-name auction websites and stores There are several marketplaces that concentrate on certain market segments and products. You can find lower prices on the items you require through these marketplaces and are more suitable to purchase bulk quantities. Manufacturers also offer their own sites on which you can order dropshipping of the items you're looking for. Items purchased directly from manufacturers are less expensive when compared to the products you can purchase from other retailers. However, certain manufacturers don't have their own stores online. They prefer selling their products through third-party sellers such as Amazon as well as Etsy. Before you decide to purchase your goods, you might need to check the costs of the top dropship companies within your vicinity. Incredibly, there are many websites that provide comparisons and contrasts of the costs of items from various online stores. There are some amazing bargains by looking at the comparison of these websites. Additionally, there are internet forums that specialize in particular items. Check these forums often to find the items they discuss. Who is to say that you won't require these items at some time? Be aware that information is valuable. A person who lives next door could need this details, but who does not know?

Be sure to trust your gut all the time. Do not just buy the product simply because you're in need of them. Think about the purchase in every kind of seriousness. If it's very good it's fine. However, if not certain be cautious! Let your instinct guide your purchasing decisions. There are many people who will offer you quick-fix schemes to get rich. If you don't take the schemes a thorough examination and you could be victimized. Naturally, a amount of money could be lost in that direction and if proper the proper precautions are not taken to avoid bankruptcy, you could be in the position of being bankrupt. In the end, you're smart or, even better you are aware of what you're looking for. Be cautious in purchasing decisionsand don't get impatient to sign any agreement. Keep in mind that life happens divided into stages, or perhaps in small steps. Begin your business step by step at one step at a time. Be sure to read past customers' opinions about the companies you plan to deal with, and what they have to have to say about the services they provide. In the end, reviews aren't just to entertain you. They're meant to help new clients to make better choices. There's no way you will make a bad choice if you go with your instincts all the time.

Buy smart

Be aware of the cost of the items you purchase. Also, think about shipping charges. A great deal might come with a high shipping cost. Be mindful of what you purchase. Consider whether the cost of shipping is worthwhile, or whether you could just go to the nearest store to purchase the item you want. Compare and contrast the prices of the different shipping options available for the product you're purchasing. If your need for the item isn't urgent, you may opt for a less expensive shipping rate. In the end, you'll save a significant amount of money by doing that. Don't rush to purchase your items through auction websites. Shipping charges on auction websites are the sole seller's discretion. They are known to increase the shipping price in an attempt to gain the trust of their clients. To reduce the cost of shipping substantially, you must purchase several products from the same seller and simultaneously. Avoid repair or refurbished items. The amount they charge, often are comparable to with new ones. While deals can appear fantastic, you'll be better off staying clear of the products in question altogether. Make sure to check the warranty in case you have to purchase refurbished items.

Check the return policies of the products you are planning to purchase. It could be necessary to return products that you bought due to one reason or other. However, this might not be feasible if you deal with dropship retailers that do provide a returns policy on its products. However, the dropship company you plan to shop with must have a complete return policy. This policy should outline clearly, what items you'll be accountable for as far as the items are involved. Some retailers may charge for restocking charges to handle returns. These charges can be deducted from the price of the product and the remaining amount is transferred to you. Find coupons. These codes will help you find amazing bargains. You can use these promo codes while purchasing your items. However, keep in mind that these codes apply to specific items. In order to make smart purchases products, look on the internet for coupons which work with the dropship store you're dealing with. You can enter the codes you discover into the fields provided and then proceed to buy your items.

Being Safe

Don't buy from a site that's not secure. Make sure that the site you shop on is secured by a padlock icon next to the address, especially when you're in the checkout procedure. This will ensure that all information you send to Amazon servers is secured. This prevents hackers from gaining access to the shared information. Therefore, if the padlock icon doesn't appear present on the ship, look for another website. Make sure to check the security of the site. every site you buy from. There should be an icon for a padlock next to the address while you are during the checkout process. It ensures that the information you enter is secure when it's transferred to Amazon servers. This blocks criminals from being able to examine the data. If you do not see the padlock icon, don't purchase from that site. Sites that take the shape of 'https://www.example.com' are good for business. But, on no occasion must you opt for websites that look like 'http://www.example.com'. Make use of a credit card to pay for your purchased products instead of the debit card. Your account is safer when you use the credit card. Why? Thieves may have direct access to your bank account when they ultimately get the data of the debit card. However, if your credit card data is stolen companies that issue credit cards will swiftly reverse the transaction. To reduce the risk it is recommended that you make use of one credit card for all online transactions. By doing this, you can determine the likelihood of data theft. Utilizing a variety of cards to make online transactions is not of any advantage since only one card can perform the task flawlessly.

Change your passwords. Use different passwords for all of your account with online stores. Keep in mind that thieves will be trying to hack into your accounts. It is easy to compromise your accounts when you use the same password across all accounts. Therefore, to stay sure make sure you change your passwords. If your accounts are compromised, your information regarding your payments could be accessed by unauthorised people, including thieves. You should also keep all receipts for your purchases. This is important in order to be able to link your expenditures to that of your account statement. Additionally, in the event of fraud, you'll be able to refer to. It is dependent on what you plan to do with the record that you have made of purchases. The receipts you receive could be printed and filed or you can have them stored digitally. Beware of malware and viruses when processing your transactions online. Your information security could be compromised by these malware and viruses. Regularly run anti-virus scans on your PC to protect yourself from these harmful viruses. Take all steps to safeguard your personal information about payments from fraudsters. But, buying products on the internet requires a few steps. In this article, we'll look at these steps one after another.

1. Make use of Google To Find The Product

Find the item you plan to purchase Check the product you want to purchase on Google or another search engine you prefer. Many websites will be displayed. Opt for the one you'd like to shop at. As you'd expect it is only after you've looked at the offerings of all stores. Additionally, you can utilize the Shopping feature of Google to find the desired products from reliable online retailers. By clicking on this option, it will let you see the products available, based on the prices they are selling at and their customer reviews. Information about the sellers and other details about the items will appear when you click on one of the items suggested. These information will aid you in deciding where to buy your dropship items. Remember that the products that are suggested were provided by a few merchants. There are many other online stores that you can shop particularly if you're unhappy with the products these merchants or sellers have to offer. When you are looking for the products you want it is also possible to take a an look at the company's website. You should ensure that the items have the same top-quality features that they have made. Repackaging is one way that some retailers may decrease the quality of the items they offer. Be on the alert!

2. Purchase the Product from a Trustworthy Website

Choose a reliable website. Don't settle for the first dropship store you stumble across randomly online. Many scammers are available and you need to be extremely cautious with your business transactions to ensure you don't get victimized. Make sure that the website you're dealing with is legitimate. Finding a reliable online store isn't difficult if you're not in a rush to buy on the internet. Also, given the chance loss of your precious cash during the process there is no other choice other than to find an authentic dropship company to manage your orders. Reviews from customers and online reviews of these stores are crucial. Additionally, you can speak with your colleagues and friends about it. Maybe, they may have purchased items online prior to. Because they're close to each other, there is a reason to believe in their judgements. But, there are many alternative online stores to consider as you go about searching for trustworthy online stores. It is possible to choose well-known stores such as Etsy, Amazon, or AliExpress in the event that you're not in a position to research all of the ways to find reputable dropshipping retailers on the internet.

Many sites will allow you to take an extensive review of the product simply by hovering your mouse over the item. Then, you are able adding the product to your shopping cart. This is only possible when you are really interested in the item. Repeat the process all over again until you've successfully included all the products you would like to purchase on the internet. Then, go straight to the checkout page. However, it important to note that opening of accounts isn't an requirement to shop in any online retailer. While some stores will need you to own an account on their platform but you do not have to open an account at certain sites before you are able to purchase their products. What you buy is entirely up to you. However, from the looks of things, buying products through your accounts is faster and easier. Naturally, you don't need to provide all of your details once more. You'd have provided that information at the time of account creation so you'll have provide your information when you purchase from an online retailer, even if there is no account. A lot of time and effort will be spent for this. So, a delay is inevitable in this business transaction.

3. Finish the Transaction

Each of the boxes marked with an asterisk must be sold prior to you finalizing the transaction. Your email, name and address could need to be verified to ensure that the products you purchased will not be delivered to an incorrect address. Be sure to provide the address and information about the person who you're buying items for specifically in the case of intended as a gift. Certain websites ship items ordered at the addresses of the buyer for the item. These issues should be taken into account before you complete the transaction online. Go to the next step after you've provided the necessary information. After that, you'll be given the chance to evaluate the purchase. Make sure that you've purchased the correct products. Why? There's nothing you're able to make once you've mastered the process of making a payment. After that, you'll need to enter the information for your debit or credit card. Verify your card information if you are happy with the product. Some verification questions could be asked. After you have provided all the answers your order will be accepted. The email you receive will inform you confirming that your order was received. Additionally, the majority of dropship sellers will send an additional email to inform you that the order is being shipped.

Chapter 11: Identifying Top-Seller Products

Once you have established your position on the marketplace, you have to determine what products you plan to offer to generate revenue for your business. One of the biggest errors that dropshippers make is filling their stores with every item that is even remotely suited to their brand. It is believed that this will increase the opportunities can be earned through your business since you have a wide range of products for customers to pick from. However, what's actually occurring is that your customers are overwhelmed by the sheer number of options, and end up finding them looking elsewhere for products they were searching for. In addition, having excessive items in your store can cause your shop to end up being sloppy because you've got more like a department store to your shop than a specialist shop. Naturally, customers prefer an exclusive shop to buy by because this creates an image that is superior in quality and consequently, is more worth making the investment.

This chapter we'll explain how you can select items that will be a success in your business. The method we're going to accomplish this is by selecting items that are focused on your particular market and are most likely to be sold. Instead of relying on just one or two sales for each of hundreds of different items to generate sales, you're relying on a greater quantity of sales by using the use of fewer products. At the conclusion, you'll see that this improves your brand's reputation and boosts sales which ultimately earns you an income that is higher more than any other approach.

Product Niches Down

In the previous chapter you determined the ideal market to build your brand to ensure that you market to a market that will be more inclined to buy from you. In this chapter, we'll utilize this particular area to help you choose the most effective products that are relevant to your particular niche and are most likely to bring you profits for your business. The products you're selecting will directly affect your image and position that you establish for yourself, which is why it is important to ensure you're selecting top quality products for your task.

When it comes to choosing what kind of product is most suitable for your brand's image and positioning, there's one aspect that you should be focusing on in each product. It is the relevance.

Each product you select to sell in your store must be in line with the niche you're in and in a very specific way. If it doesn't make sense to market it in your field, then you shouldn't be selling it in any way. It is not advisable to begin selling items which are related to neighboring niches or even to niches like yours, since it could confuse your customers as to what your specific niche is. For instance, if you're company is selling sleek, contemporary yoga equipment, you should not be selling sleek modern lunch boxes simply because customers also consume food, perhaps right after their yoga classes. This might seem like it's logical since your customers can buy yoga gear and lunch boxes they'd take their food in after their class, but to your customers, it would appear unclear. Avoid selling anything that isn't exactly what you say you'll sell. So, when visitors arrive on your website they will know precisely what you're offering and don't have to try to remember whom you're from or the products it is that you are selling to them.

Additionally, it makes it simpler for you to establish the identity of your brand and what you can offer, sticking to your area of expertise makes it simpler to promote and sell through your product, too. In the end, at all times, should your brand is famous for a specific product and market your products exclusively to the specific segment, your clients won't even be aware the other products you offer. If you decide to diversify into different areas and are not doing it as part of a strategic business plan, it is likely that your customers will not even know the fact that you have these products on offer. Therefore it is likely that they will not visit your store to purchase those items in the first place, since they won't be expecting your company to be the one that sells those items.

In the chapter you'll learn about in Chapter 11 There are situations where you could explore niches in other areas and start offering additional products to your customers. However, this should be a deliberate and planned decision that will improve your business and increase revenue in a sensible way. This shouldn't be something you attempt to accomplish immediately, before your customers have the opportunity to discover the person you are and what you can provide them.

Finding the Top Products to Sell

If you are now aware of the best way to categorize your products for your specific niche it is important to ensure that you select the top products to sell within your business. With relevance being a factor it is important to be required to ensure that you select appropriate products that you can make money with. When selecting the right products for your dropshipping business, you must ensure that you pick items that are well-known as well as profitable, priced at a reasonable price quality, durable, and easily marketable. In this article, we'll discuss how you can assess each product to ensure that it fits in these categories and helps your business to reach your desired sales figures and revenue goals.

Selecting products that are popular

The first and perhaps most obvious thing you should do to your business is to select products that are loved by your targeted market. It is important to ensure that you are selling products that people actually would like to buy. Making sure that you choose products that are popular can assist you in creating a an effective shop packed with attractive, interesting items that attract customers and inspire them to look at your store.

It is possible to determine which products are the most sought-after in your field through platforms like Amazon as well as Etsy and looking for keywords that relate to your business. By doing this, you can navigate to the relevant category for your company. It will reveal the most popular products that customers are buying. Depending on the platform you're on it is possible to alter your sorting settings to show "Top-Rated" and "Best selling." Selecting these search options will guarantee that the information being displayed to you is what customers are buying frequently not only the items that have been advertised or sponsored by the person selling the product.

When you've done a search on these platforms, you'll have a good understanding of which products are most popular for sale. After that, you can utilize platforms such as Google Trends to conduct a thorough an analysis of the product, so that you can start to determine whether it's truly popular enough to offer it in your shop. The ideal product must be able to sustain a steady uptrend in it, however it must not be in the top trend. A product that is not in the top search rankings in accordance with the parameters provided by search engines like Google Trend, is likely to be too competitive to sell in. You're looking for products that are popular and not over-saturated, so that you don't have to compete against the numerous brands offering products to the same audience.

To begin your online store You should determine approximately 30-50 items which you could offer within your shop. This might sound like an amount however it's cut down in the next steps to around 15-30 new items that you can fill your store with. This is sufficient for a brand new dropshipping business starting out and leaves the possibility of expansion over time. Therefore, do not go much more than this figure.

The Effective Pricing of Your Products

While you are looking through the top products you can sell in your store It is important to ensure that you note down the amount you can charge per product. This will aid in determining how much your margins for profit would be in the future, while providing you with an idea of how you could make use of your prices to establish yourself on the marketplace.

In terms of prices, it is important to take a look and find out what prices similar items are priced at in different stores. When you do this, try to determine the price that is lowest on the market, and the price that is the highest within the market. Next, you should determine the most well-known price point that is the one that has generated the highest sales. You may think that the lowest price is the one that generates the most sales, however consumers generally find products priced too cheap to be made cheaply. Dropshippers that offer the lowest prices possible are thought to have poor quality suppliers, leading to customers becoming skeptical about buying from their business in the first place. If you are a business it is best to set your prices about mid-point or a bit lower or more, depending on what you would like your position to be. If you wish to be viewed as cheap or even affordable, you need to ensure that you're using the lower part of the mid-point. If you wish customers to think of that you are a premium boutique shop, the cost should be a little higher. In this way, you can make your pricing a part of your position and also use it to influence how people perceive your company.

Making sure that the Profit Margins are big enough

Once you've identified the most suitable products you are able to fill your store with, you must ensure that your margins for profit are large enough. Profit margins are calculated on the amount will be required to store products in your store, against the cost you intend to market your items for. To determine your actual price , you'll need to take into account the expenses of running your company along with the price of the item itself, that you buy through your vendor.

The ideal product you offer must be able to offer at least 30% or greater profit margin to guarantee that you will earn enough revenue to pay for advertising and marketing the item and also make profits over the. Anything that is less than 30% will probably be unsuitable to sell because it won't earn you enough to be able to effectively manage your business. If you're not earning an acceptable profit margin that is sufficient to cover the cost of selling the product and selling large quantities of it, there's no need to buy anything that has an unprofitable margin.

It is essential to recognize that the more well-known your product is within the market and the more popular your product is, the lower your profit margin will to be due to an extremely competitive market your competitors. This is why it is important to choose products that are well-known and not overly saturated as it will ensure that you're getting the correct amount of money and attention on your products to achieve success with.

Finding products of high-quality to sell

Then, you must be sure you're taking into account the quality of your products. Keep in mind that even though there are many parties that are involved in making money through your business, you're the one with an image. If you've not concentrated on finding high-quality products from reputable suppliers, you may discover that your brand reputation suffers an enormous hit, and will completely hinder your potential to earn profits from dropshipping. It is important to spend time and look for suppliers who offer the finest quality of items and solutions to you clients to give you the greatest chance to build an impressive reputation and consequently, increase sales.

We'll discuss ways to find and evaluate the suppliers we will discuss in chapter 8 however, be aware that this is crucial to help you determine whether you should offer a product in your shop. If you can't discover the top quality product through a reliable supplier, you should to not stock this product at all. Instead, search until you can find the best supplier. In the meantime, you can stock the store with additional items that customers will love.

Consider How marketable a Product is

Another factor to consider is be thinking about is the extent to which a product's marketability is. When you are stocking your store with any item but if you don't find a clear method by which you can promote the product to your target market and customers, it might not be a good idea to sell that product even at all. The items you carry must be logical and be simple to integrate into your marketing plan particularly since a significant part of your strategy for marketing involves discussing and sharing about the products you offer in your company. It is important to quickly determine what you can photograph of the item and how you can integrate that image into your marketing plan, and what you can say about the product to increase sales. If you're not able to see how this all be able to work within your branding image and persona then it might not be the ideal product to sell in your shop.

Your personal experience

If you have chosen the best niche for your needs it is likely that you have an passion for the subject already, which means that you have a amount of personal experience with the subject matter, and products that fall within the particular niche. Also you must be able consult with yourself and ask questions such as "would you be interested in something similar to that?" or "would I really invest money in something similar to the one I'm looking at?" Ideally, you should stock your store with items that you can actually see yourself buying or contemplating, so that you are confident that each product you sell you sell is something others would be interested in, as well. If someone visits your shop you want them to feel that there are plenty of alternatives, not as if they're searching for one diamond in the middle of the sea. You shouldn't need to look for that only interesting item in the entire shop because your entire shop should have products that are interesting to them. This will allow them to remain for long enough to purchase something!

The Competition: Measuring

The final step in choosing the products you'll have in your store is to determine if your products are truly competitive with the other competitors. Examining the competition was an essential aspect of determining your merchandise, but today you must think about it more seriously. The most important thing to do is take a look at your competitors and find out what they're selling and then evaluate whether or not you're able to take a step forward to maintain your edge. In the ideal scenario, you will be able to discern your options to remain on top of your game by understanding the things you can provide that your competitors are not offering. Even if you're offering the same items, you need to determine the ways in which your pricing or brand can be utilized to create an experience that gives you a competitive advantage against your rivals. In this way, you will assure that your offerings are most likely to get favored over others. Particularly when you are an unproven company that is yet to establish a reputation, you'll want to ensure that you've got a competitive advantage that gives consumers a convincing reason to choose you instead of anyone else.

Chapter Summary

Your products are an essential element of your company since they offer your customers something they can purchase. If you don't have your goods, there's no way for money to flow into your company since there's nothing to purchase from you. Making the right choices with regards to your items is vital as it is your chance to ensure that you're making a positive impression for your business, and providing a great opportunity for potential customers to buy from your company.

You must ensure that your products will support your position by selecting products that fit your industry, and making sure that you select products that are able to cost effectively in your business. It is important to use the price you set as element of your positioning strategy Also, make sure that at your desired price you are still able to make a high margin of profit. This will allow you to make use of your prices and products to establish your place in the market, while earning a substantial income.

Always ensure that you have a advantage in your products so that customers prefer to shop with you instead of anyone else. The more compelling reasons you can provide to make people choose your brand over others, the more likely they will decide to shop with you instead of anyone else. This will increase the chances of success for droppingshipping your business.

Chapter 12: Instagram Analytics

You're using Instagram for business reasons and you'd like the maximum number of people to be aware of your drop-shipping business. The greater the number of followers, the greater amount of customers you'll get and the greater chance of your followers recommending your business to their friends and family. But, you need to properly manage your Instagram account in order to make yourself stand out.

You might have had the experience of logging into your personal Instagram account, and then going directly into someone else's account to check out what they shared the previous day. The same can happen for your company account, as followers will be keen to see what you've posted. Instagram analytics are a crucial tool to accomplish this.

Analytics for Instagram are vital to help you comprehend how your account performs through advertising efforts. They provide you with data that can assist you make improvements and expanding your business. In this article, we will discuss about the importance of Instagram analytics for you, and the ways to make use of them.

Why Should You Care About Instagram Analytics?

You're one of over 25 million companies with accounts on Instagram. Instagram account. According to estimates, over two hundred million users on Instagram are on at least one company profile each day. As a platform, it is a huge benefit for companies. But it is important to understand who is viewing the profile of yours, and who is viewing your posts, ads and the stories you share. Instagram analytics help you attain the following goals:

* Target the correct target.

Post frequently enough.

Draw relevant traffic to your site.

Post during optimal hours.

Take advantage of opportunities to build brand awareness.

Tak4 to take into account remarks and mentions that might be useful.

Instagram Profile Analytics

Analytics are accessible by visiting on the "My Account" page on which you publish live. The analytics will show your results over the course of a week. The metrics you can study with this tool for analysis are:

Mentions: How many number of times your website was mentioned during conversations.

Call clicks: displaying how many times the visitors on your site have reacted to your call-to-action.

Clicks on your website: Displaying the number of times followers have visited your site by clicking the link that is to your Instagram page (Miles 2014).

Visits to your profile: The number of views to your profile that you've received over the week.

Reach: An indicator of the number of people who have seen your material.

Interactions: A measurement of the actions your account has performed.

"# Impressions," a measurement of the number of times that people have been exposed to your content.

Instagram Audience Analytics

One of the most important questions these analytics aid in answering is who your target audience is. They inform you about the important demographics of your audience. The main aspects that you should consider are:

Top places You'll be able to identify the top five cities and countries in which your followers reside.

Age range Instagram breakdown the ages of your Instagram followers.

Gender: Followers are distinguished by gender, which can be seen in males and females.

The number of hours you follow: will find out the amount of time the followers you follow spends on Instagram.

Followers days: You are able to choose the times of the week on which your followers are active.

Individual Post Analytics

This is vital in determining how a message is received by your target audience. Here are a few of the data that are available for individual posts:

Commentaries: A number of comments on your posts.

#Saves: A percentage of the number of followers who are saving your blog posts.

Impressions: A gauge of the opinions of your posts.

Reach: A measurement of how many people your posts have reached.

Follows: A measurement of new followers you have received from your post.

Discovery: A feature that lets you access the latest content and posts from Instagram profiles you do not follow. You can discover the accounts you can connect to through discovery, even though they're not your followers.

Interactions: Actions that are taken directly from your blog like using the email or call, hyperlink, or even logging into your profile.

Instagram Stories Analytics

It is crucial to continue to post Instagram stories. They give you the opportunities to not only let people know about your company but also increase the popularity of your business. They allow you to build the brand's image to new customers and boost exposure to your existing audience. Analytic tools can monitor the following:

Insights.

Replies.

173

Exits The amount of times someone quit your story at any point in the story.

Reach.

Impressions.

The main goal you must have when tracking analytics is creating brand awareness. Instagram is the ideal platform to promote brand awareness, particularly for new companies since Instagram boasts more than one billion users. The metrics that can help you with this include:

* * Follower counts: A rise in followers is an increase in brand recognition. To boost brand awareness you could participate in joint campaigns with the right partners, create advertisements, or hold contests to make it easy for people to follow and follow your content.

* Impressions: It's all about making sure your account, story and post impressions are increasing as this can increase your brand's recognition.

"Reach": This lets you keep track of the number of views for your page, your

174

stories as well as posts. With reach, you'll be capable of seeing your users' individual accounts. Expanding your reach is essential for brand recognition.

• Sales a variety of methods you can employ to boost sales via Instagram including posting offers to your followers or running specific advertisements.

Chapter 13: Psychological As Well As Technical Scenarios For Scaling Up Your Business

The role of technology and psychology is an important role in the success and development of your future eCommerce business. There are tested methods that will help your company to increase profits. It is important to think about and answer honestly about what your business requires to grow. Knowing the problems and difficulties you face in running your business can assist in establishing a plan to fix the problems.

Knowing what your business requires in relation to solutions is the first step to scaling up. The next step is to study the things you've learned about your company, which will help you determine the most effective technological and/or psychological situation to take into consideration.

The importance of psychology is in the field of sales and marketing. The patterns of people's spending and their behavior have been examined with confirmed results on triggers that cause a consumer to buy items. This is the reason why people's habits of spending, online behavior such as social media websites they frequent, as well as their purchases have been thoroughly studied to create these concepts in both technical and psychological marketing. Here are some examples of situations that you can apply to increase selling and customer retention.

Color psychology: Colors have significant influence on our tastes or dislikes. If your website employs certain hues that cause an emotion in your visitors is psychological in nature. Colors also affect gender and other aspects. An example of a reaction to color could be that of the hue (red) that stimulates and relaxes, as well as the color (green) that relaxes.

This type of psychological model should be used when creating your site. Choose the colors you will use to attract the right type of people who will be visiting your site in the future. If you're just beginning to design your website and decide on the colors you will use, take a look out other websites that are in your field, and see what attracts you to their offerings, and how it relates to the items themselves.

Recency illusion: With this psychological notion, it's comprised of the art of repetition advertising campaign. That is, it shows customers the product in a large amount. For customers, it appears that at every turn they are greeted by ads to their products. This is a good thing and an effective method to boost sales.

Utilize the ego of your customer - Make your customers feel welcomed and appreciated every time they visit your shop. If people feel that they are unique to your company, they're likely towards being loyal patrons that can convey their feelings to other customers and encourage business to visit your site.

A few ways to show your appreciation could be to give them a discount to customers who visit your shop. Offers of freebies are always appreciated. inform them that customers can download, for instance an eBook for free just because you are grateful for your business. Happy customers will be back as potential buyers So keep them interested and reap the benefits.

Be rational . All people don't require reasons to buy items. Certain people prefer honest communications. Offering information about the product and how it can meet the need or desire together with the details of the product is all you need. Certain people perceive manipulative psychology, and it might deter them from using your website , or from becoming a customer.

Always be honest and open when you are presenting your product or service. Reviewing the product can benefit those who are hesitant to buy because they come from other people who have experienced the product and have first-hand experience with.

To create excitement, make the most of a new item to attract attention and create anticipation. Although the product might not be as significant a big as you want it to appear, the customers' mind will be excited and will want to purchase this particular product. Consider spinners as an example.

There was a lot of conversation about the product and word of mouth was spread around about the product, making it an immediate success. There was no way to get the item in stock because they sold out quickly. They were quite popular in the early days and consisted of a spinning toy which was extremely inexpensive to create. But the hype it generated was what made it a huge success.

Zendesk is a software that aids in providing high-quality customer service. It gathers information from your social media channels regarding any concerns of customers. Zendesk collects all customer interaction via email, chat, and even search, and puts them all in one location to be analysed to determine the most effective customer service solutions.

As entrepreneurs, you may not have time to respond to all customer inquiries or, if you're a newly-established business, you might not have the money to pay for an entire team of people to assist customers. Zendesk is perfect for this situation. Zendesk can also send out surveys to customers.

Conclusion

Dropshipping is a model of business that doesn't require an inventory inventory prior to the start of your business. Dropshipping is simple. You study products to determine which products you would like to dropship. When you've decided on which product to offer then you locate a business which offers dropshipping services for the specific product. Once you have decided what you want to offer, take a look at the prices of the company and add your profits to the cost. The price you pay will be the amount that your customers will pay. Once you have determined the cost, you develop a website or an ad and send customers to your site using the advertisement. When the customer has placed the purchase on your site after which you go back to the website that dropsships the item and place the order of the customer there. You are responsible for the cost and take the profits. The dropshipping business then ships items to your customers directly from their warehouse, meaning you do not have to handle the item at all. By just only a couple of clicks of an icon, you are able to keep the profits without having to deal with inventory or completing orders.

Skills in research are essential for this model of business. It is essential to find drop-shippers who dropship your product. You should also look through the reviews on the business to determine what other customers have experienced good experiences with the business. If you have found the company you'd like to work for You are now ready to act. Additionally, you must be proficient in writing for this model of business. You'll have to write content for your site to explain what the product does. Additionally, you'll need to be able to write ads.

When people type words into Google's Google search bar it is because they're using keyword phrases. Being able to identify these keywords and then use to your advantage is the core of the ability to research keywords. By entering keywords into Google that your prospective customers might be searching for helps you determine whether the product you intend to market is something customers would like to purchase. Knowing you have terms that consumers are looking for will make you more successful. One of the best places to search for keywords is to use this tool called the Google Keyword Planner, which is a no-cost tool. SEO is a term used to describe search engine optimization. these abilities can help you ensure that your site is to the top of Google. Understanding how to use SEO can assist you in bringing the sales you want. There are specific places where you can include words to help give your site an extra boost. If you don't have the SEO expertise you need now but it's not something you should worry about. It is possible to slowly master and improve your website in the future.

www.ingramcontent.com/pod-product-compliance
Lightning Source LLC
Chambersburg PA
CBHW071219210326
41597CB00016B/1870